Version 13

Quality and Process Methods

"The real voyage of discovery consists not in seeking new landscapes, but in having new eyes."

Marcel Proust

JMP, A Business Unit of SAS
SAS Campus Drive
Cary, NC 27513

The correct bibliographic citation for this manual is as follows: SAS Institute Inc. 2016. *JMP® 13 Quality and Process Methods*. Cary, NC: SAS Institute Inc.

JMP® 13 Quality and Process Methods

ISBN 978-1-62960-480-0 (Hardcopy)
ISBN 978-1-62960-584-5 (EPUB)
ISBN 978-1-62960-585-2 (MOBI)

SAS Institute Inc., SAS Campus Drive, Cary, North Carolina 27513-2414.

September 2016

Contents

Quality and Process Methods

13 Cause-and-Effect Diagrams

A References

Index

Chapter 1

Learn about JMP

Documentation and Additional Resources

This chapter includes the following information:

- book conventions
- JMP documentation
- JMP Help
- additional resources, such as the following:
  - other JMP documentation
  - tutorials
  - indexes
  - Web resources
  - technical support options

Formatting Conventions

The following conventions help you relate written material to information that you see on your screen:

- Sample data table names, column names, pathnames, filenames, file extensions, and folders appear in Helvetica font.
- Code appears in `Lucida Sans Typewriter` font.
- Code output appears in *`Lucida Sans Typewriter`* italic font and is indented farther than the preceding code.
- **Helvetica bold** formatting indicates items that you select to complete a task:
  - buttons
  - check boxes
  - commands
  - list names that are selectable
  - menus
  - options
  - tab names
  - text boxes
- The following items appear in italics:
  - words or phrases that are important or have definitions specific to JMP
  - book titles
  - variables
  - script output
- Features that are for JMP Pro only are noted with the JMP Pro icon JMP PRO. For an overview of JMP Pro features, visit http://www.jmp.com/software/pro/.

Note: Special information and limitations appear within a Note.

Tip: Helpful information appears within a Tip.

JMP Documentation

JMP offers documentation in various formats, from print books and Portable Document Format (PDF) to electronic books (e-books).

- Open the PDF versions from the **Help > Books** menu.
- All books are also combined into one PDF file, called *JMP Documentation Library*, for convenient searching. Open the *JMP Documentation Library* PDF file from the **Help > Books** menu.
- You can also purchase printed documentation and e-books on the SAS website:

  http://www.sas.com/store/search.ep?keyWords=JMP

JMP Documentation Library

The following table describes the purpose and content of each book in the JMP library.

Document Title	Document Purpose	Document Content
Discovering JMP	If you are not familiar with JMP, start here.	Introduces you to JMP and gets you started creating and analyzing data.
Using JMP	Learn about JMP data tables and how to perform basic operations.	Covers general JMP concepts and features that span across all of JMP, including importing data, modifying columns properties, sorting data, and connecting to SAS.
Basic Analysis	Perform basic analysis using this document.	Describes these Analyze menu platforms: • Distribution • Fit Y by X • Tabulate • Text Explorer Covers how to perform bivariate, one-way ANOVA, and contingency analyses through Analyze > Fit Y by X. How to approximate sampling distributions using bootstrapping and how to perform parametric resampling with the Simulate platform are also included.

Document Title	Document Purpose	Document Content
Essential Graphing	Find the ideal graph for your data.	Describes these Graph menu platforms: • Graph Builder • Overlay Plot • Scatterplot 3D • Contour Plot • Bubble Plot • Parallel Plot • Cell Plot • Treemap • Scatterplot Matrix • Ternary Plot • Chart The book also covers how to create background and custom maps.
Profilers	Learn how to use interactive profiling tools, which enable you to view cross-sections of any response surface.	Covers all profilers listed in the Graph menu. Analyzing noise factors is included along with running simulations using random inputs.
Design of Experiments Guide	Learn how to design experiments and determine appropriate sample sizes.	Covers all topics in the DOE menu and the Specialized DOE Models menu item in the Analyze > Specialized Modeling menu.

Document Title	Document Purpose	Document Content
Fitting Linear Models	Learn about Fit Model platform and many of its personalities.	Describes these personalities, all available within the Analyze menu Fit Model platform: • Standard Least Squares • Stepwise • Generalized Regression • Mixed Model • MANOVA • Loglinear Variance • Nominal Logistic • Ordinal Logistic • Generalized Linear Model

<table>
<tr><th>Document Title</th><th>Document Purpose</th><th>Document Content</th></tr>
<tr><td>Predictive and Specialized Modeling</td><td>Learn about additional modeling techniques.</td><td>Describes these Analyze > Predictive Modeling menu platforms:<ul><li>Modeling Utilities</li><li>Neural</li><li>Partition</li><li>Bootstrap Forest</li><li>Boosted Tree</li><li>K Nearest Neighbors</li><li>Naive Bayes</li><li>Model Comparison</li><li>Formula Depot</li></ul>Describes these Analyze > Specialized Modeling menu platforms:<ul><li>Fit Curve</li><li>Nonlinear</li><li>Gaussian Process</li><li>Time Series</li><li>Matched Pairs</li></ul>Describes these Analyze > Screening menu platforms:<ul><li>Response Screening</li><li>Process Screening</li><li>Predictor Screening</li><li>Association Analysis</li></ul>The platforms in the Analyze > Specialized Modeling > Specialized DOE Models menu are described in Design of Experiments Guide.</td></tr>
</table>

Document Title	Document Purpose	Document Content
Multivariate Methods	Read about techniques for analyzing several variables simultaneously.	Describes these Analyze > Multivariate Methods menu platforms: • Multivariate • Principal Components • Discriminant • Partial Least Squares Describes these Analyze > Clustering menu platforms: • Hierarchical Cluster • K Means Cluster • Normal Mixtures • Latent Class Analysis • Cluster Variables
Quality and Process Methods	Read about tools for evaluating and improving processes.	Describes these Analyze > Quality and Process menu platforms: • Control Chart Builder and individual control charts • Measurement Systems Analysis • Variability / Attribute Gauge Charts • Process Capability • Pareto Plot • Diagram

Document Title	Document Purpose	Document Content
Reliability and Survival Methods	Learn to evaluate and improve reliability in a product or system and analyze survival data for people and products.	Describes these Analyze > Reliability and Survival menu platforms: • Life Distribution • Fit Life by X • Cumulative Damage • Recurrence Analysis • Degradation and Destructive Degradation • Reliability Forecast • Reliability Growth • Reliability Block Diagram • Repairable Systems Simulation • Survival • Fit Parametric Survival • Fit Proportional Hazards
Consumer Research	Learn about methods for studying consumer preferences and using that insight to create better products and services.	Describes these Analyze > Consumer Research menu platforms: • Categorical • Multiple Correspondence Analysis • Multidimensional Scaling • Factor Analysis • Choice • MaxDiff • Uplift • Item Analysis
Scripting Guide	Learn about taking advantage of the powerful JMP Scripting Language (JSL).	Covers a variety of topics, such as writing and debugging scripts, manipulating data tables, constructing display boxes, and creating JMP applications.

Document Title	Document Purpose	Document Content
JSL Syntax Reference	Read about many JSL functions on functions and their arguments, and messages that you send to objects and display boxes.	Includes syntax, examples, and notes for JSL commands.

Note: The **Books** menu also contains two reference cards that can be printed: The *Menu Card* describes JMP menus, and the *Quick Reference* describes JMP keyboard shortcuts.

JMP Help

JMP Help is an abbreviated version of the documentation library that provides targeted information. You can open JMP Help in several ways:

- On Windows, press the F1 key to open the Help system window.
- Get help on a specific part of a data table or report window. Select the Help tool from the **Tools** menu and then click anywhere in a data table or report window to see the Help for that area.
- Within a JMP window, click the **Help** button.
- Search and view JMP Help on Windows using the **Help > Help Contents**, **Search Help**, and **Help Index** options. On Mac, select **Help > JMP Help**.
- Search the Help at http://jmp.com/support/help/ (English only).

Additional Resources for Learning JMP

In addition to JMP documentation and JMP Help, you can also learn about JMP using the following resources:

- Tutorials (see “Tutorials” on page 26)
- Sample data (see “Sample Data Tables” on page 26)
- Indexes (see “Learn about Statistical and JSL Terms” on page 26)
- Tip of the Day (see “Learn JMP Tips and Tricks” on page 26)
- Web resources (see “JMP User Community” on page 27)
- JMPer Cable technical publication (see “JMPer Cable” on page 27)
- Books about JMP (see “JMP Books by Users” on page 28)
- JMP Starter (see “The JMP Starter Window” on page 28)

- Teaching Resources (see "Sample Data Tables" on page 26)

Tutorials

You can access JMP tutorials by selecting **Help > Tutorials**. The first item on the **Tutorials** menu is **Tutorials Directory**. This opens a new window with all the tutorials grouped by category.

If you are not familiar with JMP, then start with the **Beginners Tutorial**. It steps you through the JMP interface and explains the basics of using JMP.

The rest of the tutorials help you with specific aspects of JMP, such as designing an experiment and comparing a sample mean to a constant.

Sample Data Tables

All of the examples in the JMP documentation suite use sample data. Select **Help > Sample Data Library** to open the sample data directory.

To view an alphabetized list of sample data tables or view sample data within categories, select **Help > Sample Data**.

Sample data tables are installed in the following directory:

On Windows: C:\Program Files\SAS\JMP\13\Samples\Data

On Macintosh: \Library\Application Support\JMP\13\Samples\Data

In JMP Pro, sample data is installed in the JMPPRO (rather than JMP) directory. In JMP Shrinkwrap, sample data is installed in the JMPSW directory.

To view examples using sample data, select **Help > Sample Data** and navigate to the Teaching Resources section. To learn more about the teaching resources, visit http://jmp.com/tools.

Learn about Statistical and JSL Terms

The **Help** menu contains the following indexes:

Statistics Index Provides definitions of statistical terms.

Scripting Index Lets you search for information about JSL functions, objects, and display boxes. You can also edit and run sample scripts from the Scripting Index.

Learn JMP Tips and Tricks

When you first start JMP, you see the Tip of the Day window. This window provides tips for using JMP.

To turn off the Tip of the Day, clear the **Show tips at startup** check box. To view it again, select **Help > Tip of the Day.** Or, you can turn it off using the Preferences window. See the *Using JMP* book for details.

Tooltips

JMP provides descriptive tooltips when you place your cursor over items, such as the following:

- Menu or toolbar options
- Labels in graphs
- Text results in the report window (move your cursor in a circle to reveal)
- Files or windows in the Home Window
- Code in the Script Editor

Tip: On Windows, you can hide tooltips in the JMP Preferences. Select **File > Preferences > General** and then deselect **Show menu tips.** This option is not available on Macintosh.

JMP User Community

The JMP User Community provides a range of options to help you learn more about JMP and connect with other JMP users. The learning library of one-page guides, tutorials, and demos is a good place to start. And you can continue your education by registering for a variety of JMP training courses.

Other resources include a discussion forum, sample data and script file exchange, webcasts, and social networking groups.

To access JMP resources on the website, select **Help > JMP User Community** or visit https://community.jmp.com/.

JMPer Cable

The JMPer Cable is a yearly technical publication targeted to users of JMP. The JMPer Cable is available on the JMP website:

http://www.jmp.com/about/newsletters/jmpercable/

JMP Books by Users

Additional books about using JMP that are written by JMP users are available on the JMP website:

http://www.jmp.com/en_us/software/books.html

The JMP Starter Window

The JMP Starter window is a good place to begin if you are not familiar with JMP or data analysis. Options are categorized and described, and you launch them by clicking a button. The JMP Starter window covers many of the options found in the Analyze, Graph, Tables, and File menus. The window also lists JMP Pro features and platforms.

- To open the JMP Starter window, select **View** (**Window** on the Macintosh) **> JMP Starter**.
- To display the JMP Starter automatically when you open JMP on Windows, select **File > Preferences > General**, and then select **JMP Starter** from the Initial JMP Window list. On Macintosh, select **JMP > Preferences > Initial JMP Starter Window**.

Technical Support

JMP technical support is provided by statisticians and engineers educated in SAS and JMP, many of whom have graduate degrees in statistics or other technical disciplines.

Many technical support options are provided at http://www.jmp.com/support, including the technical support phone number.

Chapter 2

Introduction to Quality and Process Methods

Tools for Process and Product Improvement

This book describes a number of methods and tools that are available in JMP to help you evaluate and improve quality and process performance:

- Control charts provide feedback on key variables and show when a process is in, or out of, statistical control. Chapter 3, "Control Chart Builder" and Chapter 4, "Shewhart Control Charts" describe the JMP approach to creating control charts, including an interactive control chart platform called Control Chart Builder. When you need to detect smaller shifts in a process or monitor multiple process characteristics simultaneously, see Chapter 5, "Cumulative Sum Control Charts" and Chapter 6, "Multivariate Control Charts" respectively.
- The Measurement Systems Analysis platform assesses the precision, consistency, and bias of a system. Before you can study a process, you need to make sure that you can accurately and precisely measure the process. If variation comes from the process itself, then you are not reliably learning about the process. Use this analysis to find out how your system is performing. For more information, see Chapter 7, "Measurement Systems Analysis".
- The Variability/Attribute Gauge Chart platform creates variability or attribute gauge charts. Variability charts analyze continuous measurements and reveal how your system is performing. Attribute charts analyze categorical measurements and show you measures of agreement across responses. You can also perform a gauge study to see measures of variation in your data. For more information, see Chapter 8, "Variability Gauge Charts" and Chapter 9, "Attribute Gauge Charts" respectively.
- The Process Capability platform measures the ability of a process to meet specification limits. You can compare process performance, summarized by process centering and variability, to specification limits. The platform calculates capability indices based on both long-term and short-term variation. The analysis helps identify the variation relative to the specifications; this enables you to achieve increasingly higher conformance values. For more information, see Chapter 10, "Process Capability".
- The Capability Analysis platform, found in the Distribution platform, determines the ability of a process to meet specification limits. The platform enables you to compare a process to specific tolerances. For more information, see Chapter 11, "Capability Analysis".
- The Pareto Plot platform shows the frequency of problems in a quality related process or operation. Pareto plots help you decide which problems to solve first by highlighting the frequency and severity of problems. For more information, see Chapter 12, "Pareto Plots".

- The Diagram platform constructs cause-and-effect diagrams, which organize the sources of a problem for brainstorming or as a preliminary analysis to identify variables for further experimentation. Once complete, further analysis can be done to identify the root cause of the problem. For more information, see Chapter 13, "Cause-and-Effect Diagrams".

Figure 2.1 Quality and Process Methods Examples

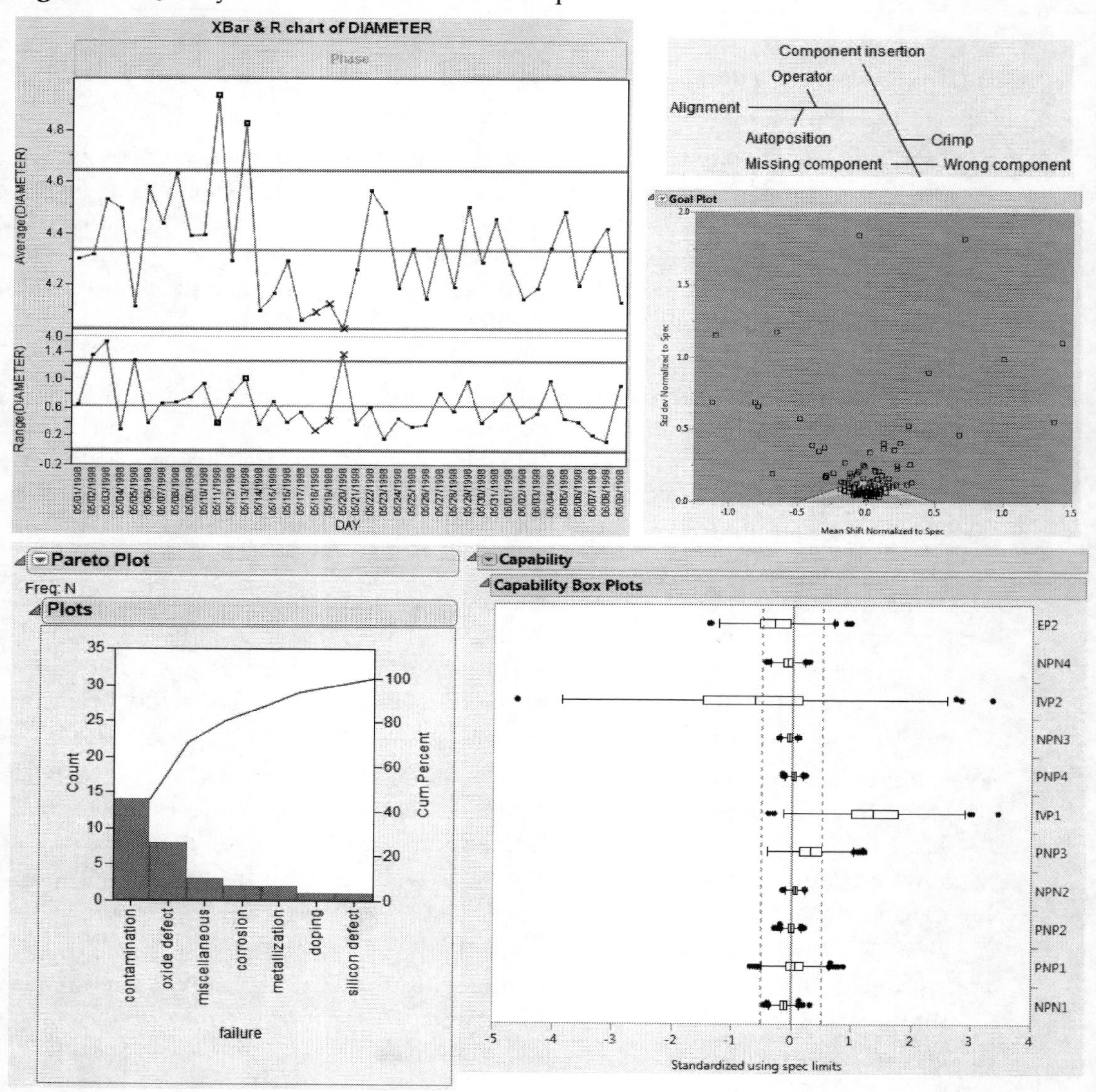

Chapter 3

Control Chart Builder

Create Control Charts Interactively

A control chart is a graphical and analytic tool for monitoring process variation. Use Control Chart Builder to create control charts of your process data. Select the variables that you want to chart and drag them into zones. JMP automatically chooses the appropriate chart type based on the data. The instant feedback encourages further exploration of the data. You can change your mind and quickly create another type of chart, or change the current settings on the existing chart.

Figure 3.1 Control Chart Builder Example

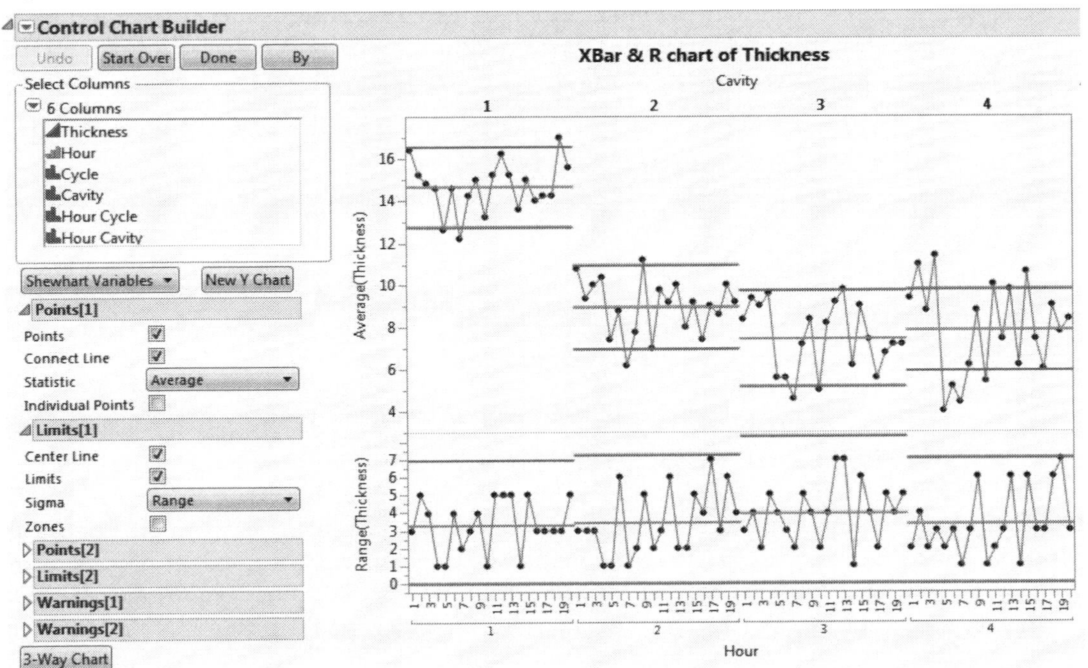

Overview of the Control Chart Builder

A control chart is a graphical way to filter out routine variation in a process. Filtering out routine variation helps manufacturers and other businesses determine whether a process is stable and predictable. If the variation is more than routine, the process can be adjusted to create higher quality output at a lower cost.

This version of JMP continues a shift in the approach to control charts. We are moving toward an all-in-one, interactive workspace called the Control Chart Builder. The Control Chart Builder enables you to create several types of control charts (Shewhart Variables, Shewhart Attribute, and Rare Event) and is intended to be an interactive tool for problem solving and process analysis. Shewhart control charts are broadly classified into control charts for variables and control charts for attributes. Rare event charts are useful for events that occur so infrequently that a traditional chart is inappropriate.

To use Control Chart Builder, you do not need to know the name of a particular chart beforehand. When you drag a data column to the workspace, Control Chart Builder creates an appropriate chart based on the data type and sample size. Once the basic chart is created, use the menus and other options to:

- Change the type of chart. You can switch between Attribute, Variables, and Rare Event charts without relaunching the platform.
- Change the statistic on the chart. You can add, remove, and switch variables without relaunching the platform.
- Format the chart and create subgroups that are defined by multiple *X* variables.
- Add additional charts, including three-in-one charts: subgroup means, within-subgroup variation, and between-subgroup variation.

Example of the Control Chart Builder

This example uses the Socket Thickness.jmp sample data table, which includes measurements for the thickness of sockets. There has been an increase in the number of defects during production and you want to investigate why this is occurring. Use Control Chart Builder to investigate the variability in the data and the control of the process.

1. Select **Help > Sample Data Library** and open Quality Control/Socket Thickness.jmp.
2. Select **Analyze > Quality and Process > Control Chart Builder**.
3. Drag Thickness to the **Y** zone.
4. Drag Hour to the **Subgroup** zone (at bottom).

Figure 3.2 Control Charts for Socket Thickness

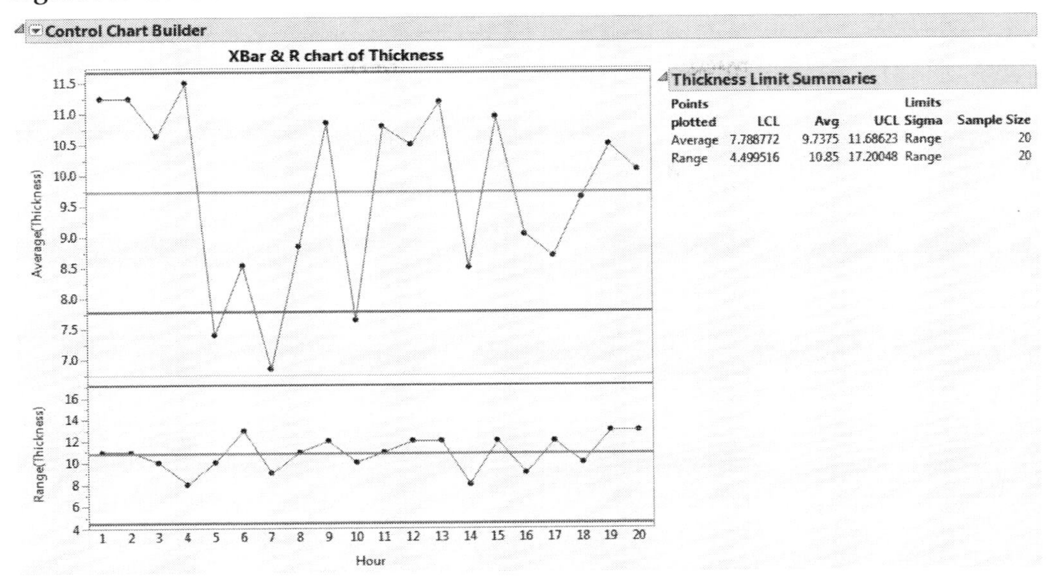

Looking at the Average chart, you can see that there are several points below the lower control limit of 7.788772. You want to see whether another variable might be contributing to the problem.

5. Drag **Cavity** into the **Phase** zone.

Figure 3.3 Control Charts for Each Cavity

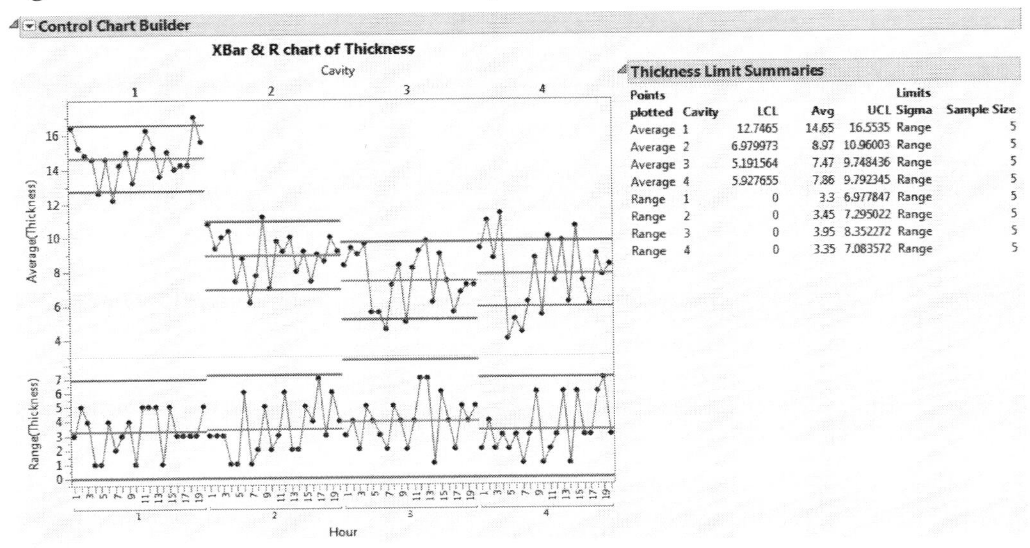

From the Average chart, you can conclude the following:

- There are differences between the cavities, each deserving separate control limits.
- Cavity 1 is producing sockets with a higher average thickness, indicating that further investigation of the differences between cavities is warranted.
- All of the cavities have points that are outside the control limits. Therefore, you should investigate the lack of control in the data for each cavity.

The Range chart for each cavity shows that the within-subgroup measurements are in control.

Control Chart Types

The Control Chart Builder (CCB) enables you to create several types of control charts (Shewhart Variables, Shewhart Attribute, and Rare Event). To create a chart, you do not need to know the name or structure of a particular chart beforehand. Select the variables (or columns) that you want to chart, and then drag and drop them into zones. When you drag a data column to the workspace, Control Chart Builder creates an appropriate chart based on the data type and sample size. Once the basic chart is created, you can use the menus and other options to change the type, the statistic, and the format of the chart.

Shewhart Control Charts for Variables

Control charts for variables are classified according to the subgroup summary statistic plotted on the chart.

- $\overline{X}$-charts display subgroup means (averages)
- *R*-charts display subgroup ranges (maximum – minimum)
- *S*-charts display subgroup standard deviations
- Presummarize charts display subgroup means and standard deviations
- Individual Measurement charts display individual measurements
- Moving Range charts display moving ranges of two successive measurements

XBar-, R-, and S- Charts

For quality characteristics measured on a continuous scale, a typical analysis shows both the process mean and its variability with a mean chart aligned above its corresponding *R*- or *S*-chart.

Individual Measurement Charts

Individual Measurement charts displays individual measurements. Individual Measurement charts are appropriate when only one measurement is available for each subgroup sample. If you are charting individual measurements, the individual measurement chart shows above its corresponding moving range chart. Moving Range charts displays moving ranges of two successive measurements.

Presummarize Charts

If your data consist of repeated measurements of the same process unit, you can combine these into one measurement for the unit. Pre-summarizing is not recommended unless the data have repeated measurements on each process or measurement unit.

Presummarize summarizes the process column into sample means and/or standard deviations, based either on the sample size or sample label chosen. Then it charts the summarized data based on the options chosen in the window.

Levey-Jennings Charts

Levey-Jennings charts show a process mean with control limits based on a long-term sigma. The control limits are placed at 3*s* distance from the center line. The standard deviation, *s*, for the Levey-Jennings chart is calculated the same way standard deviation is in the Distribution platform.

Shewhart Control Charts for Attributes

In the previous types of charts, measurement data was the process variable. This data is often continuous, and the charts are based on theory for continuous data. Another type of data is count data or level counts of character data, where the variable of interest is a discrete count of the number of defects or blemishes per subgroup. For discrete count data, attribute charts are applicable, as they are based on binomial and Poisson models. Because the counts are measured per subgroup, it is important when comparing charts to determine whether you have a similar number of items in the subgroups between the charts. Attribute charts, like variables charts, are classified according to the subgroup sample statistic plotted on the chart.

Table 3.1 Attribute Chart Determination

	Statistic	
Sigma	**Proportion**	**Count**
Binomial	p-chart	np-chart
Poisson	u-chart	c-chart

The CCB makes a few decisions for you based on the variable selected. For example, if there is no *X* variable, a c-chart is originally created because there is no way to estimate the binomial distributions. Upon adding an *X* variable (or lot size), the platform switches to a np-chart if the count per subgroup is less than the subgroup sample size. Once the basic chart is created, you can use the menus and other options to change the type, the statistic, and the format of the chart.

- p-charts display the proportion of nonconforming (defective) items in subgroup samples, which can vary in size. Because each subgroup for a p-chart consists of N_i items, and an item is judged as either conforming or nonconforming, the maximum number of nonconforming items in a subgroup is N_i.
- np-charts display the number of nonconforming (defective) items in subgroup samples. Because each subgroup for a np-chart consists of N_i items, and an item is judged as either conforming or nonconforming, the maximum number of nonconforming items in subgroup i is N_i.
- c-charts display the number of nonconformities (defects) in a subgroup sample that usually, but does not necessarily, consists of one inspection unit.
- u-charts display the number of nonconformities (defects) per unit in subgroup samples that can have a varying number of inspection units.

Rare Event Control Charts

A Rare Event chart is a control chart that provides information about a process where the data comes from rarely occurring events. Tracking processes that occur infrequently on a traditional control chart tend to be ineffective. Rare event charts were developed in response to the limitations of control charts in rare event scenarios. The Control Chart Builder provides two types of rare event charts (g- and t-charts).

A g-chart is used to count the number of events between rarely occurring errors or nonconforming incidents, and creates a chart of a process over time. Each point represents the number of units between occurrences of a relatively rare event. For example, in a production setting, where an item is produced daily, an unexpected line shutdown can occur. You can use a g-chart to look at the number of units produced between line shutdowns. A traditional plot of data such as this is not conducive to control chart interpretation. The g-chart helps visualize such data in traditional control chart form.

A t-chart measures the time elapsed since the last event and creates a picture of a process over time. Each point on the chart represents an amount of time that has passed since a prior occurrence of a rare event. A traditional plot of this data might contain many points at zero and an occasional point at one. A t-chart avoids flagging numerous points as out of control. The t-chart helps identify special and common cause variation, so that appropriate improvements can be made.

A t-chart can be used for numeric, nonnegative data, date/time data, and time-between data:

- Numeric, nonnegative data is the number of intervals between events. It can be continuous or integer.

- Date/time data records the date and time of each event. Each data value must be greater than or equal to the preceding value.
- Time-between data (also known as elapsed-time data) represent the elapsed time between event *i* and event *i-1*.

Like the g-chart, the t-chart is used to detect changes in the rate at which the adverse event occurs. When reading the t-chart, the points above the upper control limit indicate that the amount of time between events has increased. Thus, the rate of the events has decreased. Points below the lower control limit indicate that the rate of adverse events has increased.

Because of how time is measured for these charts, one fundamental difference is that a point flagged as out of control above the limits is generally considered a desirable effect because it represents a significant increase in the time between events. The difference between a g- and t-chart is the scale used to measure distance between events. The g-chart uses a discrete scale, whereas the t-chart uses a continuous scale.

Table 3.2 Rare Event Chart Determination

	Statistic
Sigma	**Count**
Negative Binomial	g-chart
Weibull	t-chart

Control Chart Types

The most common control charts are available in the Control Chart Builder and in the Control Chart platform. Use the Control Chart Builder as your first choice to easily and quickly generate charts. JMP automatically chooses the appropriate chart type based on the data. Table 3.3 through Table 3.8 summarize the different control chart types.

Table 3.3 Variable Charts Without Grouping (X) Variable or Nonsummarized Data

Chart Types	**Control Chart Builder Options**	
	Points > Statistic	**Limits > Sigma**
Individual	Individual	Moving Range
Moving Range on Individual	Moving Range	Moving Range
Levey Jennings	Individual	Levey Jennings

Table 3.4 Variable Charts with Grouping (X) Variables or Summarized Data

Chart Types	Control Chart Builder Options	
	Points > Statistic	**Limits > Sigma**
XBar (limits computed on range)	Average	Range
XBar (limits computed on standard deviation)	Average	Standard Deviation
R	Range	Range
S	Standard Deviation	Standard Deviation
Levey Jennings	Individual measurements. Control limits are based on an estimate of long-term sigma.	Levey Jennings or overall Standard Deviation

Table 3.5 Presummarize Charts

Chart Types	Control Chart Builder Options	
	Points > Statistic	**Limits > Sigma**
Individual on Group Means	Average	Moving Range
Individual on Group Std Devs	Standard Deviation	Moving Range
Moving Range on Group Means	Moving Range on Means	Moving Range
Moving Range on Group Std Devs	Moving Range on Std Dev	Moving Range

Table 3.6 Attribute Charts

Chart Types	Control Chart Builder Options	
	Points > Statistic	**Limits > Sigma**
p-chart	Proportion	Binomial
np-chart	Count	Binomial
c-chart	Count	Poisson
u-chart	Proportion	Poisson

Table 3.7 Attribute Charts

Chart Types	Control Chart Builder Options	
	Points > Statistic	Limits > Sigma
p-chart	Proportion	Binomial
np-chart	Count	Binomial
c-chart	Count	Poisson
u-chart	Proportion	Poisson

Table 3.8 Rare Event Charts

Chart Types	Control Chart Builder Options	
	Points > Statistic	Limits > Sigma
g-chart	Count	Negative Binomial
t-chart	Count	Weibull

Launch the Control Chart Builder

Launch the Control Chart Builder by selecting **Analyze > Quality and Process > Control Chart Builder.**

Figure 3.4 Initial Control Chart Builder Window

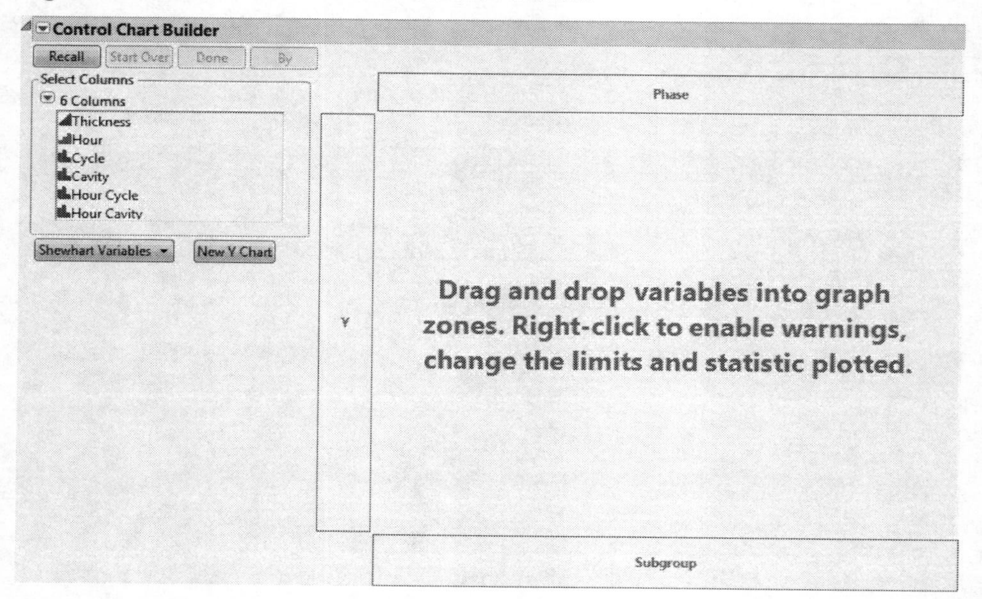

To begin creating a control chart, drag variables from the **Select Columns** box into the zones. If you drop variables in the center, JMP guesses where to put them based on whether the variables are continuous or categorical. The Control Chart Builder contains the following zones:

Y Assigns the process variable.

Subgroup Assigns subgroup variables. To define subgroup levels as a combination of multiple columns, add multiple variables to the **Subgroup** zone. When a subgroup variable is assigned, each point on the control chart corresponds to a summary statistic for all of the points in the subgroup.

Phase Assigns phase variables. When a **Phase** variable is assigned, separate control limits are computed for each phase. See also "Add Color to Delineate Phases" on page 57

The initial Control Chart Builder window contains the following buttons:

Recall Populates the window with the last analysis that you performed. The **Recall** button becomes the **Undo** button once you perform an action.

Undo Reverses the last change made to the window.

Start Over Returns the window to the default condition, removing all data, and clearing all zones.

Done Hides the buttons and the **Select Columns** box and removes all drop zone outlines. In this presentation-friendly format, you can copy the graph to other programs. To restore the window to the interactive mode, click **Show Control Panel** on the Control Chart Builder red triangle menu.

By Identifies the variable and produces a separate analysis for each value that appears in the column.

Shewhart Variables/Shewhart Attribute/Rare Event Allows you to select Shewhart Variables, Shewhart Attribute, or Rare Event control chart types. If you select an Attribute chart type, an **n Trials** box and zone appear on the chart.

n Trials Assigns a lot size when an attribute control chart is selected. Appears if you select an Attribute chart type.

New Y Chart Produces a copy of the current chart for every column selected in the **Select Columns** box. The new charts use the selected columns in the **Y** role.

Once you drag variables to the chart, other buttons and options appear at the left bottom of the screen that enable you to show, hide or switch items on the chart (See Figure 3.5). Many of these functions (Points, Limits, Warnings) are the same as the functions available when you right-click the chart. For more information, refer to "Options Panel and Right-Click Chart Options" on page 45. For information about warnings and rules, see "Tests" on page 48 and "Westgard Rules" on page 51.

3-way Chart Enables you to produce a three-way chart for variable chart types. The subgroup size must be greater than one. The plotting statistic is based on subgroup averages, within-subgroup variation, or between-subgroup variation. The default set of three includes a presummarized chart of the averages using Moving Range limits, a Moving Range chart and a Range chart.

Event Chooser Allows the chart to respond in real time to selection changes. There are several standard groups of responses that are recognized and pre-scored (for example, pass/fail, yes/no, Likert Scales, conforming/non-conforming, and defective/non-defective). If you are analyzing results from a survey and want to focus solely on a specific sector of the results for one or more questions, you can make the selection on the screen and the chart rescores and replots the chart immediately. The **Event Chooser** is available for attribute charts with response columns that have a modeling type of nominal or ordinal. The Event Chooser does not appear for response columns with a modeling type of continuous.

The Control Chart Builder Window

The analysis produces a chart that can be used to determine whether a process is in a state of statistical control. The report varies depending on which type of chart you select. Control charts update dynamically as data is added or changed in the data table. Figure 3.5 displays the Control Chart Builder window for the Bottle Tops.jmp sample data table.

To create the chart:

1. Select **Help > Sample Data Library** and open Quality Control/Bottle Tops.jmp.
2. Select **Analyze > Quality and Process > Control Chart Builder**.
3. Drag Status to the **Y** zone.
4. Drag Sample to the **Subgroup** zone.

Figure 3.5 Control Chart Builder Window

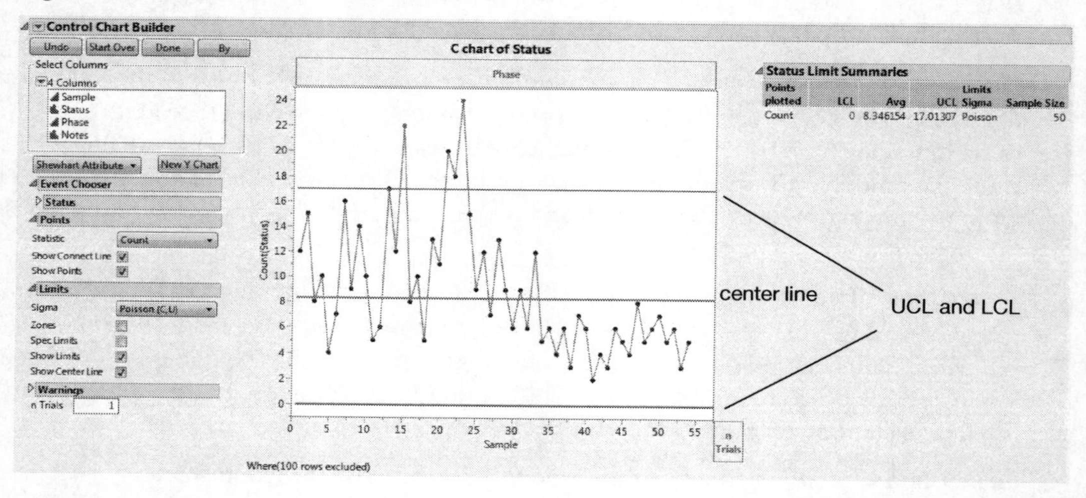

You can drag other variables into the various zones to augment the analysis and use the "Control Chart Builder Options" to further examine the data. Some of the right-click chart options (for example, show or hide points, limits, warnings, and zones; select statistic and sigma options) also appear on the left hand side of the chart for easy access.

Control charts have the following characteristics:

- Each point plotted on the chart represents an individual process measurement or summary statistic. Subgroups should be chosen *rationally*, that is, they should be chosen to maximize the probability of seeing a true process signal *between* subgroups.
- The vertical axis of a control chart is scaled in the same units as the summary statistic.

- The horizontal axis of a control chart identifies the subgroup samples and is time ordered. Observing the process over time is important in assessing if the process is changing.

The green line is the center line, or the average of the data. The center line indicates the average (expected) value of the summary statistic when the process is in statistical control. Measurements should appear equally on both sides of the center line. If not, this is possible evidence that the process average is changing.

- The two red lines are the upper and lower control limits, labeled UCL and LCL. These limits give the range of variation to be expected in the summary statistic when the process is in statistical control. If the process is exhibiting only routine variation, then all the points should fall randomly in that range.
- A point outside the control limits signals the presence of a special cause of variation.

Options in the Control Chart Builder window create control charts that can be updated dynamically as samples are received and recorded or added to the data table. When a control chart signals abnormal variation, action should be taken to return the process to a state of statistical control if the process degraded. If the abnormal variation indicates an improvement in the process, the causes of the variation should be studied and implemented.

When you double-click the axes, the appropriate Axis Specification window appears for you to specify the format, axis values, number of ticks, gridline, reference lines, and other options to display.

Control Chart Builder Options

Control Chart Builder options appear in the red triangle menu or by right-clicking on a chart or axis. Some of the right-click chart options also appear on the bottom left hand side of the chart for easy access. You can also set preferences for many of the options in the Control Chart Builder at **File > Preferences > Platforms > Control Chart Builder**.

Red Triangle Menu Options

Show Control Panel shows or hides the following elements:

- buttons
- the **Select Columns** box
- the drop zone borders

Show Limit Summaries Shows or hides the Limit Summaries report. This report shows the control limits (LCL and UCL), the center line (Avg), the Points and Limits plotted, and the Sample Size for the chart. Sample size is not shown for rare event charts.

Show Capability (Available only for Variable plots that have a column with a Spec Limits column property.) Shows or hides the Process Capability Analysis report. For more information about the Process Capability Analysis report, see "The Process Capability Report" on page 230.

Get Limits Retrieves the control limits that are stored in a data table.

Set Sample Size Sets a subgroup size. Missing values are taken into account when computing limits and sigma.

Save Limits Saves the control limits as a column property in the existing data table for the response variable. The option does not work with phase variables.

Save Summaries Creates a new data table containing such information as the sample label, sample sizes, statistic being plotted, center line, control limits, and any tests, warnings and failures. The specific statistics included in the table depend on the type of chart.

Include Missing Categories Enables the graph to collect rows with missing values in a categorical column, and displays the missing values on the graph as a separate category. If this option is disabled, all rows with a missing X value are removed from the calculations, in addition to being hidden from the graph.

This option is not available for continuous X variables or categorical Y variables because there is no compelling way to display the collected missing values on the relevant axes. By default, this option is enabled.

Note: If Include Missing Categories is enabled, capability analysis results in Control Chart Builder do not match those in the Process Capability platform if a categorical X variable has missing values.

See the JMP Reports chapter in the *Using JMP* book for more information about the following options:

Local Data Filter Shows or hides the local data filter that enables you to filter the data used in a specific report.

Redo Contains options that enable you to repeat or relaunch the analysis. In platforms that support the feature, the Automatic Recalc option immediately reflects the changes that you make to the data table in the corresponding report window.

Save Script Contains options that enable you to save a script that reproduces the report to several destinations.

Save By-Group Script Contains options that enable you to save a script that reproduces the platform report for all levels of a By variable to several destinations. Available only when a By variable is specified in the launch window.

Note: Column Switcher is available only for a single Y variable having two or fewer associated charts. Based on the selected chart type, only columns that are appropriate for the Y role are included in the Column Switcher column list.

Note: In Control Chart Builder, the Automatic Recalc option is turned on by default and cannot be turned off.

Options Panel and Right-Click Chart Options

The following options appear on the left hand side of the chart for easy access and when you right-click a chart.

Points Provides the following options:

- **Statistic** changes the statistic plotted on the chart. See "Statistic" on page 46.
- **Individual Points** show or hides individual observations in a subgroup. Available only with a subgroup variable or Set Sample Size. This option is not available for Attribute chart types or Rare Event charts.
- **Show Connect Line** shows connecting lines between the points.
- **Show Points** shows or hides the points on the chart.

Limits Provides the following options:

- **Sigma** specifies the method of computing sigma. See "Sigma" on page 47.
- **Zones** shows or hides the zones on the chart. The zones are defined as one, two, and three sigmas on either side of the mean. Control Chart Builder does not extend the size of one zone over another. If the limits are not centered around the mean, (UCL-Avg)/3 is used as the width of each zone. Zones are not drawn below the LCL or above the UCL. Available only for Variables and Attribute chart types.
- **Spec Limits** shows or hides the specification limits on the chart. Appears only if the data table has a Spec Limits column property. The Column Info Window chapter in the *Using JMP* book includes details about adding this column property.
- **Set Control Limits** enables you to enter control limits for tests. After you click **OK** in the Set Control Limits window, the specified control limits are set uniformly across groups. Select this option again to remove the specified control limits.
- **Add Limits** specifies additional control limits to be plotted on the chart. These limits are not used in tests.
- **Show Limits** hides or shows the control limits on the chart.
- **Show Center Line** hides or shows the center line on the chart.

Add Dispersion Chart Adds a dispersion chart to the chart area. Change the chart type with the **Points** options. A dispersion chart illustrates the variation in the data by plotting one of many forms of dispersion, including the range, standard deviation, or moving range. Available only for Variables chart types.

Set Sample Size Sets a subgroup size. Missing values are taken into account when computing limits and sigma.

Warnings Provides the following options:

– **Customize Tests** lets you design custom tests and select or deselect multiple tests at once. After the option is selected, the Customize Tests window appears for designing the tests. Select a test description, and enter the desired number (n) and label. You can save the settings to preferences and also restore the default settings. Available only for Variables and Attribute chart types.

– **Tests** let you select which statistical control tests to enable. For more information about tests, see "Tests" on page 48. Available only for Variables and Attribute chart types.

Note: Move your cursor over a flagged point on the chart to see a description of the test that failed.

– **Westgard Rules** lets you select which Westgard statistical control tests to enable. Because Westgard rules are based on sigma and not the zones, they can be computed without regard to constant sample size. For more information about tests, see "Westgard Rules" on page 51. Available only for Variables and Attribute chart types.

– **Test Beyond Limits** enables the test for any points beyond the control limits. These points are identified on the chart. This test works on all charts with limits, regardless of the sample size being equal.

Remove Graph Removes the control chart. Available on the second and subsequent control charts in an analysis that has multiple Y charts.

Note: For a description of the **Rows**, **Graph**, **Customize**, and **Edit** menus, see the *Using JMP* book.

Statistic

You can change the statistic represented by the points on the chart. The options available depend on the chart type selected.

For Variables chart types, you can change the statistic represented by the points on the chart using the following options:

Individual Creates a chart where each point represents an individual value in the data table.

Average Creates a chart where each point represents the average of the values in a subgroup.

Range Creates a chart where each point represents the range of the values in a subgroup.

Standard Deviation Creates a chart where each point represents the standard deviation of the values in a subgroup.

Moving Range on Means Computes the difference in the range between two consecutive subgroup means.

Moving Range on Std Dev Computes the difference in the range between two consecutive subgroup standard deviations.

Moving Range Creates a chart where each point is the difference between two consecutive observations.

Note: The Average, Range, Standard Deviation, Moving Range on Means, and Moving Range on Std Dev methods appear only if a subgroup variable with a sample size greater than one is specified or a sample size is set.

For Attribute chart types, you can change the statistic represented by the points on the chart using the following options:

Proportion Creates a chart where each point represents the proportion of items in subgroup samples.

Count Creates a chart where each point represents the number of items in subgroup samples.

For Rare Event chart types, the statistic represented by the points on the chart uses the following option:

Count Creates a chart where each point represents the number of items in subgroup samples.

Sigma

You can change the method for computing sigma for the chart. The options available depend on the chart type selected.

For Variables chart types, you can use the following options:

Range Uses the range of the data in a subgroup to estimate sigma.

Standard Deviation Uses the standard deviation of the data in a subgroup to estimate sigma.

Moving Range Uses the moving ranges to estimate sigma. The moving range is the difference between two consecutive points.

Levey-Jennings Uses the standard deviation of all the observations to estimate sigma.

For Attribute chart types, you can use the following options:

Binomial Uses the binomial distribution model to estimate sigma. The model indicates the number of successes in a sequence of experiments, each of which yields success with some probability. Selecting Binomial yields either a p- or np-chart.

Poisson Uses the Poisson distribution model to estimate sigma. The model indicates the number of events and the time at which these events occur in a given time interval. Selecting Poisson yields either a c- or u-chart.

For Rare Event chart types, you can use the following options:

Negative Binomial Uses the negative binomial distribution model to estimate sigma. The model indicates the number of successes in a sequence of trials before a specified number of failures occur. Selecting Negative Binomial yields a g-chart.

Weibull Uses the Weibull distribution model to estimate sigma. The model indicates the mean time between failures. Selecting Weibull yields a t-chart.

Tests

The **Warnings** option in the right-click menu or on the left hand side of the window displays a submenu for **Tests** selection. You can select one or more tests for special causes (Western Electric rules) from the menu. Nelson (1984) developed the numbering notation used to identify special tests on control charts. The tests work with both equal and unequal sample sizes.

If a selected test is positive for a particular sample, that point is labeled with the test number. When you select several tests for display and more than one test signals at a particular point, the label of the numerically lowest test specified appears beside the point. You can move your cursor over a flagged point on the chart to see a description of the test that failed.

Tip: To add or remove several tests at once, select or deselect the tests in the Control Panel under **Warnings > Tests**.

Table 3.9 on page 49 lists and interprets the eight tests, and Figure 3.7 illustrates the tests. The following rules apply to each test:

- The area between the upper and lower limits is divided into six zones, each with a width of one standard deviation.
- The zones are labeled A, B, C, C, B, A with zones C nearest the center line.
- A point lies in Zone B or beyond if it lies beyond the line separating zones C and B. That is, if it is more than one standard deviation from the center line.
- Any point lying on a line separating two zones lines is considered belonging to the innermost zone. So, if a point lies on the line between Zone A and Zone B, the point is considered to be in Zone B.

Tests 1 through 8 apply to all Shewhart chart types.

Tests 1, 2, 5, and 6 apply to the upper and lower halves of the chart separately. Tests 3, 4, 7, and 8 apply to the whole chart.

See Nelson (1984, 1985) for further recommendations on how to use these tests.

Figure 3.6 Zones for Western Electric Rules

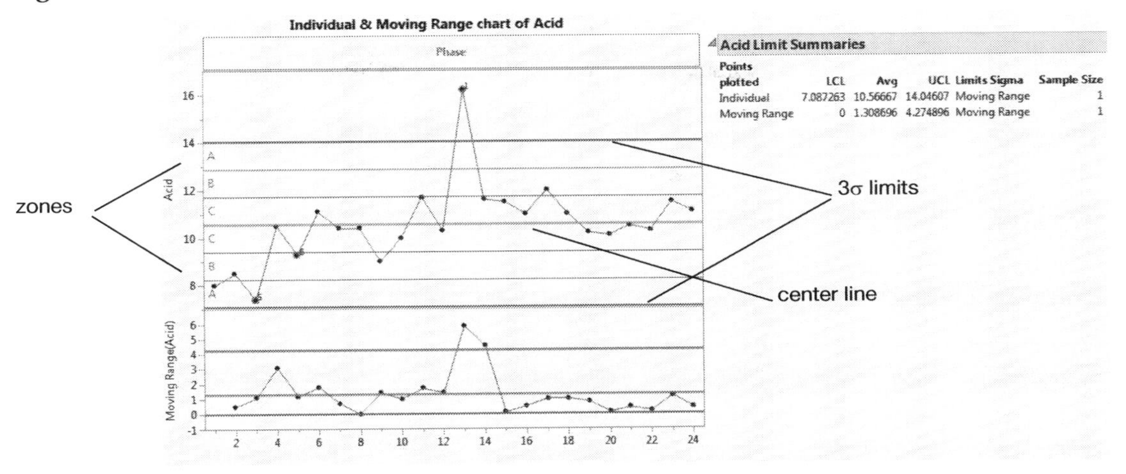

Table 3.9 Description and Interpretation of Tests for Special Causes[a]

Test 1	One point beyond Zone A	Detects a shift in the mean, an increase in the standard deviation, or a single aberration in the process. For interpreting Test 1, any dispersion chart (*R*-, *S*-, or *MR*-) can be used to rule out increases in variation.
Test 2	Nine points in a row in a single (upper or lower) side of Zone C or beyond	Detects a shift in the process mean.
Test 3	Six points in a row steadily increasing or decreasing	Detects a trend or drift in the process mean. Small trends are signaled by this test before Test 1.
Test 4	Fourteen points in a row alternating up and down	Detects systematic effects such as two alternately used machines, vendors, or operators.
Test 5	Two out of three points in a row in Zone A or beyond and the point itself is in Zone A or beyond.	Detects a shift in the process average or increase in the standard deviation. Any two out of three points provide a positive test.

Table 3.9 Description and Interpretation of Tests for Special Causes[a] *(Continued)*

Test 6	Four out of five points in a row in Zone B or beyond and the point itself is in Zone B or beyond.	Detects a shift in the process mean. Any four out of five points provide a positive test.
Test 7	Fifteen points in a row in Zone C, above and below the center line	Detects stratification of subgroups when the observations in a single subgroup come from various sources with different means.
Test 8	Eight points in a row on both sides of the center line with none in Zones C	Detects stratification of subgroups when the observations in one subgroup come from a single source, but subgroups come from different sources with different means.

a. Nelson (1984, 1985)

Figure 3.7 Illustration of Special Causes Tests[1]

Test 1: One point beyond Zone A

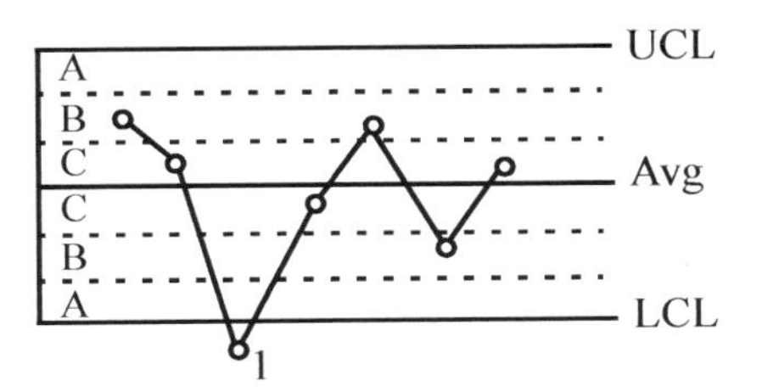

Test 2: Nine points in a row in a single (upper or lower) side of Zone C or beyond

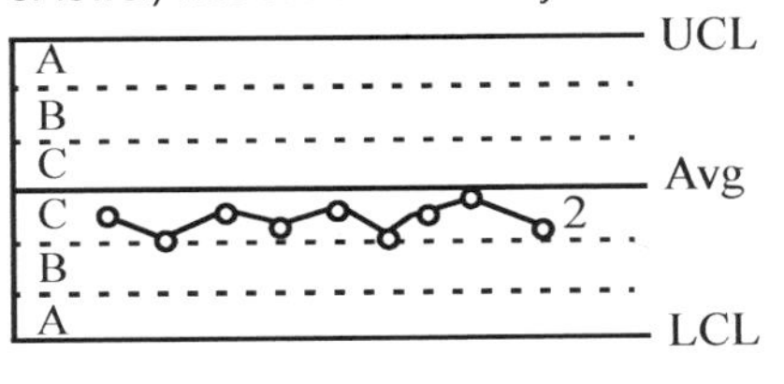

Test 3: Six points in a row steadily increasing or decreasing

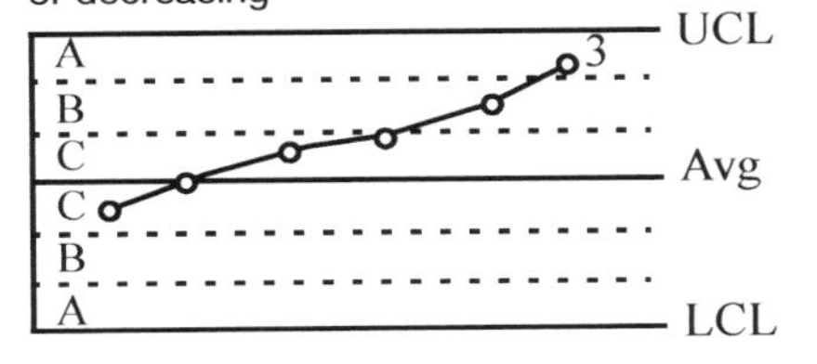

Test 4: Fourteen points in a row alternating up and down

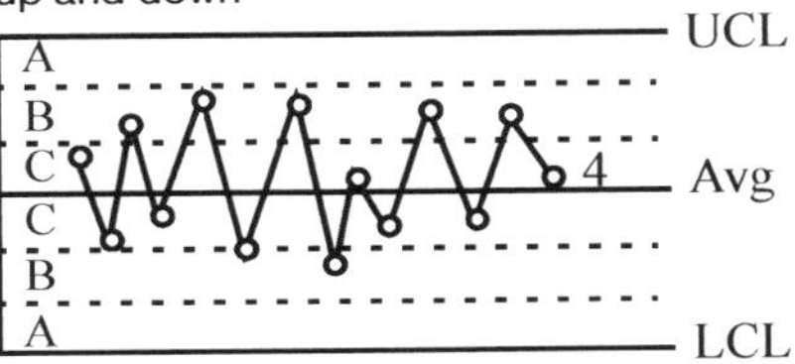

Test 5: Two out of three points in a row in Zone A or beyond

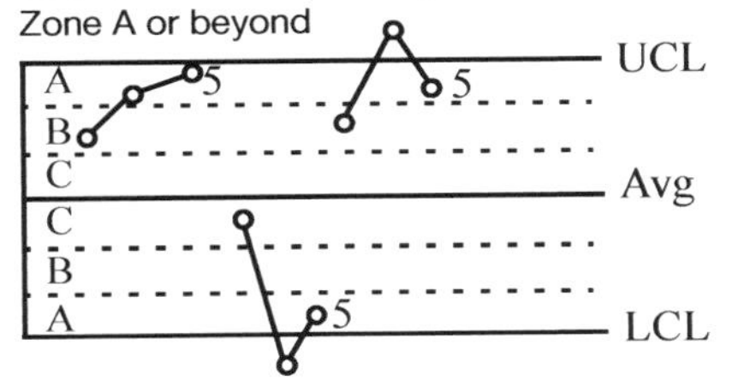

Test 6: Four out of five points in a row in Zone B or beyond

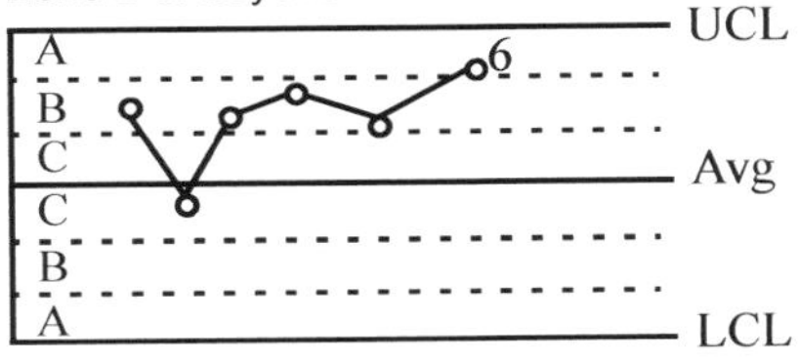

Test 7: Fifteen points in a row in Zone C (above and below the center line)

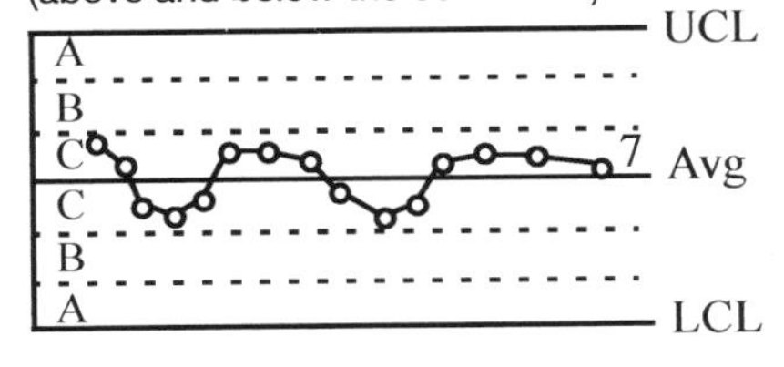

Test 8: Eight points in a row on both sides of the center line with none in Zone C

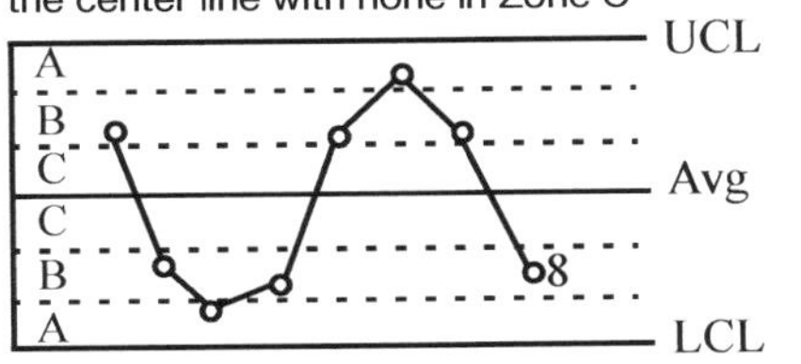

Westgard Rules

Westgard rules are implemented under the **Westgard Rules** submenu of the **Warnings** option when you right-click on a chart or on the left hand side of the window. The different tests are abbreviated with the decision rule for the particular test. For example, **1 2s** refers to a test where one point is two standard deviations away from the mean.

---

1. Nelson (1984, 1985)

Rule 1 2S is commonly used with Levey-Jennings charts, where control limits are set 2 standard deviations away from the mean. The rule is triggered when any one point goes beyond these limits.

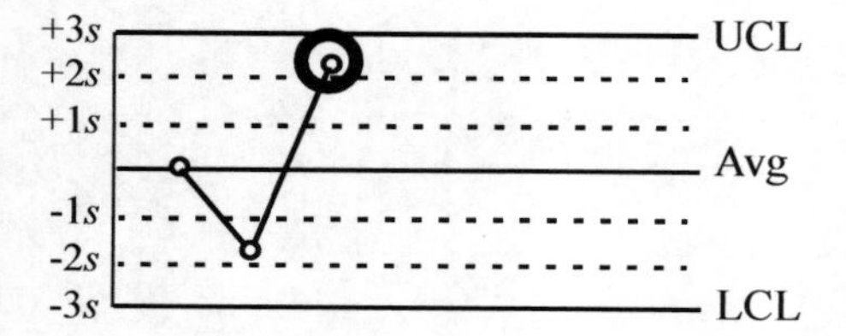

Rule 1 3S refers to a rule common to Levey-Jennings charts where the control limits are set 3 standard deviations away from the mean. The rule is triggered when any one point goes beyond these limits.

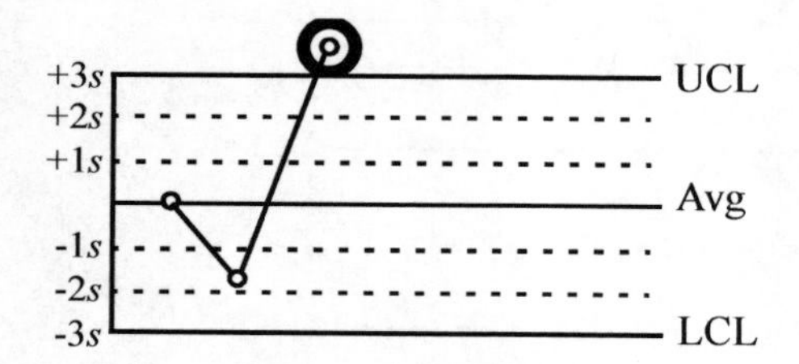

Rule 2 2S is triggered when two consecutive control measurements are farther than two standard deviations from the mean.

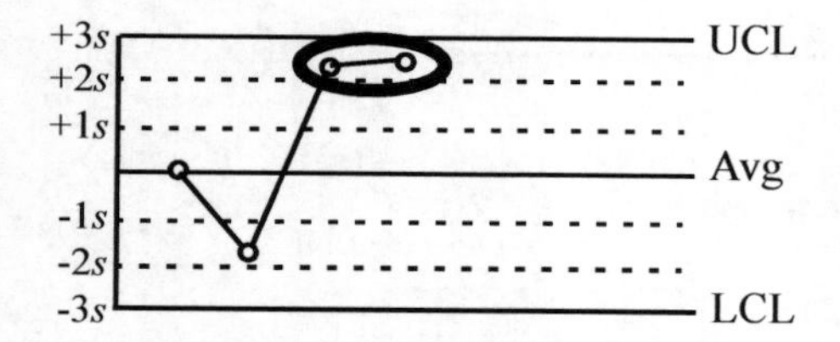

Rule R 4S is triggered when one measurement is greater than two standard deviations from the mean and the previous measurement is greater than two standard deviations from the mean in the opposite direction such that the difference is greater than 4 standard deviations.

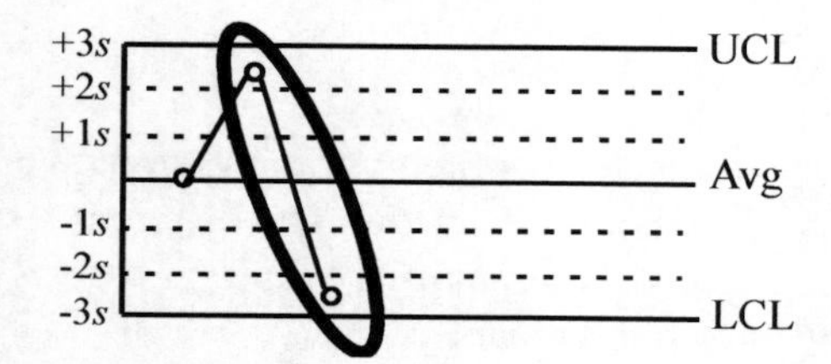

Rule 4 1S is triggered when four consecutive measurements are more than one standard deviation from the mean.

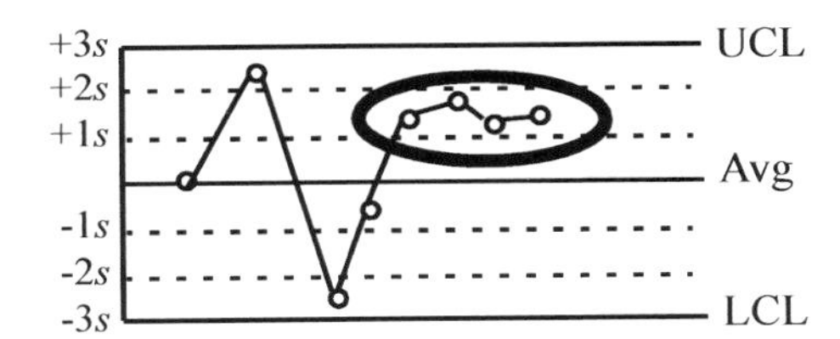

Rule 10 X is triggered when ten consecutive points are on one side of the mean.

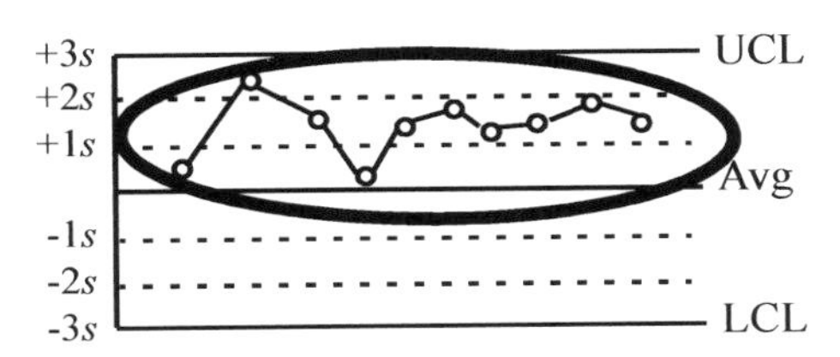

Right-Click Axis Options

Remove Removes a variable.

For details about the Axis Settings, Revert Axis, Add or Remove Axis Label, and Edit options, see the JMP Reports chapter in the *Using JMP* book.

Retrieving Limits from a Data Table

JMP can use previously established control limits for control charts:

- Upper and lower control limits, and a center line value.
- Parameters for computing limits such as a mean and standard deviation.

The control limits or limit parameter values can be either in a JMP data table, referred to as the *Limits Table*, or stored as a column property in the process column. You can retrieve the Limits Table with the **Get Limits** option on the Control Chart Builder red triangle menu.

All Limits Tables must have:

- A column of special keywords that identify each row.
- A column for each of the variables whose values are the known standard parameters or limits. This column name must be the same as the corresponding process variable name in the data table to be analyzed by the Control Chart Builder.

The Control Chart Builder identifies the appropriate limits from keywords in the _LimitsKey column. A list of limit keywords and their associated control chart is shown in Table 3.10.

Note the following:

- Rows with unknown keywords and rows marked with the excluded row state are ignored.
- Except for _Sample Size, any needed values not specified are estimated from the data.

Table 3.10 Limits Table Keys with Appropriate Charts and Meanings

Keywords	For Charts	Meaning
_KSigma	All *except* Control Chart Builder	multiples of the standard deviation of the statistics to calculate the control limits; set to missing if the limits are in terms of the alpha level
_Alpha	All *except* Control Chart Builder	Type I error probability used to calculate the control limits; used if multiple of the standard deviation is not specified in the CCB window or in the Limits Table
_Std Dev	$\overline{X}$-, *R*-, *S*-, IM, MR	known process standard deviation
_U	*c*-, *u*-	known average number of nonconformities per unit
_P	*np*-, *p*-	known value of average proportion nonconforming
_LCL, _UCL	$\overline{X}$-, IM, *p*-, *np*-, *c*-, *u*-, *g*-, *t*-	lower and upper control limit for Mean Chart, Individual Measurement chart, or any attribute or rare event chart
_AvgR	*R*-, MR	average range or average moving range
_LCLR, _UCLR	*R*-, MR	lower control limit for *R*- or MR chart upper control limit for *R*- or MR chart
_AvgS, _LCLS, _UCLS	*S*-Chart	average standard deviation, upper and lower control limits for *S*-chart

Table 3.10 Limits Table Keys with Appropriate Charts and Meanings *(Continued)*

Keywords	For Charts	Meaning
_AvgR_PreMeans _AvgR_PreStdDev _LCLR_PreMeans _LCLR_PreStdDev _UCLR_PreMeans _UCLR_PreStdDev _Avg_PreMeans _Avg_PreStdDev _LCL_PreMeans _LCL_PreStdDev _UCL_PreMeans _UCL_PreStdDev	IM, MR	Mean, upper, and lower control limits based on pre-summarized group means or standard deviations.

In the Control Chart Builder red triangle menu, you can save limits as a data table column property. To save limits as a new data table, use the Control Chart platform. See "Saving and Retrieving Limits" on page 93 for details.

Excluded and Hidden Samples

The following bullets summarize the effects of various conditions on samples and subgroups:

- Excluded subgroups are not used in the calculations, but appear in the chart (although dimmed).
- Hidden observations are used in the calculations, but do not appear in the chart.
- Both hidden and excluded rows are included in the count of points for Tests for Special Causes. An excluded row can be labeled with a special cause flag. A hidden point cannot be labeled. If the flag for a Tests for Special Causes is on a hidden point, it will not appear in the chart.
- For partially excluded subgroups, if one or more observations within a subgroup is excluded, and at least one observation within the subgroup is included, the excluded observation is not included in the calculations of either the point statistic or the limits.
- Checks for negative and non-integer data happen on the entire data (even excluded values).

- Tests continue to apply to all excluded subgroups. Excluded samples are flagged when tests are turned on.

Additional Examples of the Control Chart Builder

The following are additional examples of the Control Chart Builder. Some examples show the Control Panel while others do not. To show or hide the Control Panel, select Show Control Panel from the red triangle menu.

$\overline{X}$ and R Chart Phase Example

A manufacturer of medical tubing collected tube diameter data for a new prototype. The data was collected over the past 40 days of production. After the first 20 days (phase 1), some adjustments were made to the manufacturing equipment. Analyze the data to determine whether the past 20 days (phase 2) of production are in a state of control.

1. Select **Help > Sample Data Library** and open Quality Control/Diameter.jmp.
2. Select **Analyze > Quality and Process > Control Chart Builder**.
3. Drag DIAMETER to the **Y** role.
4. Drag DAY to the **Subgroup** role.

Figure 3.8 Control Charts for Diameter

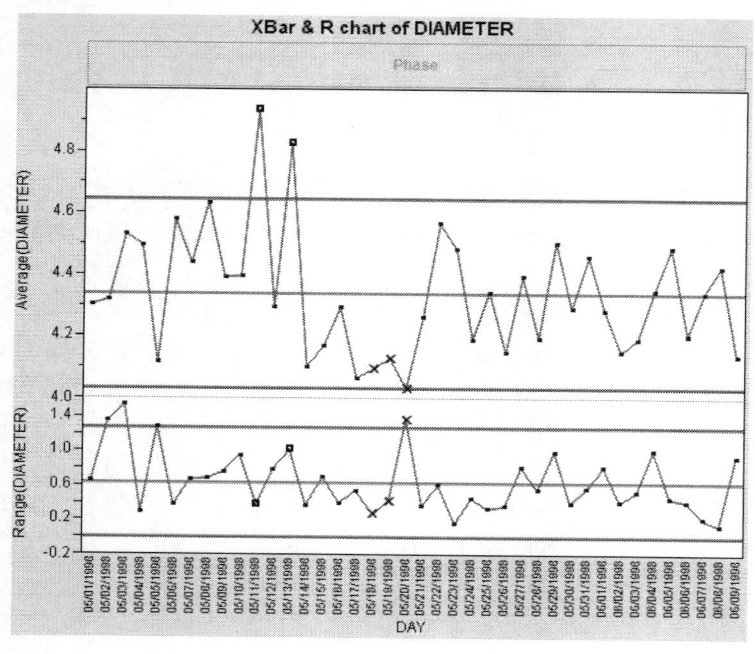

The first 20 days appear to have high variability, and in the Average chart, there are three observations that are outside of the control limits. An adjustment was made to the manufacturing equipment and new control limits were incorporated.

To compute separate control limits for each phase:

5. Drag **Phase** to the **Phase** role.
6. In the Average chart, right-click and select **Warnings > Test Beyond Limits.**

Figure 3.9 Control Charts for each Phase

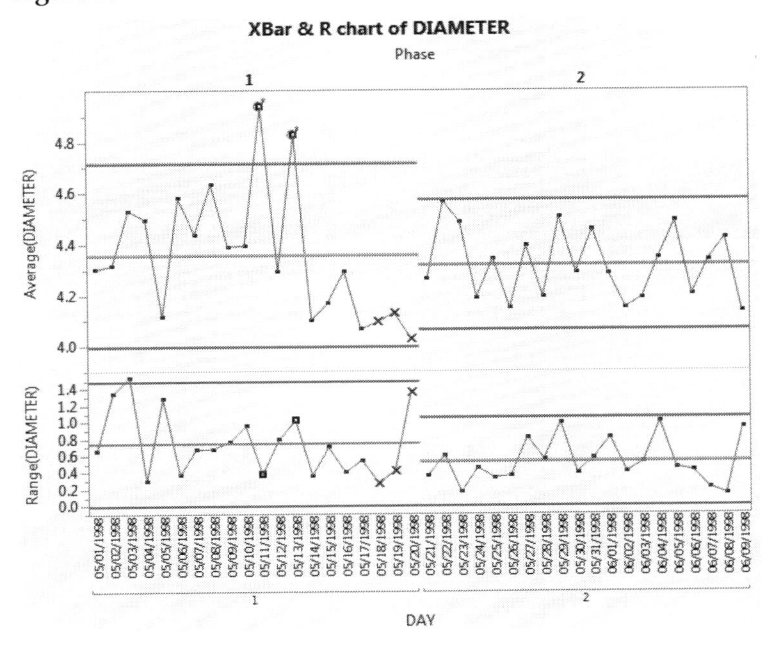

Including the **Phase** variable means that the control limits for phase 2 are based only on the data for phase 2. None of the phase 2 observations are outside the control limits. Therefore, you can conclude that the process is in control after the adjustments were made.

Add Color to Delineate Phases

If you have distinct phases in your control chart, you can illustrate them by adding different background colors to the different phases.

1. Starting from Figure 3.9, double-click in the X axis.

   The Axis Settings window appears. In the Reference Lines panel, notice that there is an existing line reference value at 19.5. This value is the midpoint of the range for DAY and also happens to be the dividing value between the two phases.

2. Select **Allow Ranges**.

3. Enter -0.5 for the **Min Value** (the scale minimum).
4. Enter 19.5 for the **Max Value** (the dividing line).
5. Choose a color, say yellow. Change the opacity to 40%.
6. Click **Add.**
7. Click **Allow Ranges.**
8. Enter 19.5 for the **Min Value** (the dividing line).
9. Enter 39.5 for the **Max Value** (the maximum of the axis).
10. Choose a color, say light blue. Change the opacity to 40%.
11. Click **Add.**

    You can see from the preview how the chart will look.
12. Click **OK.**

Figure 3.10 Diameter Phases with Color

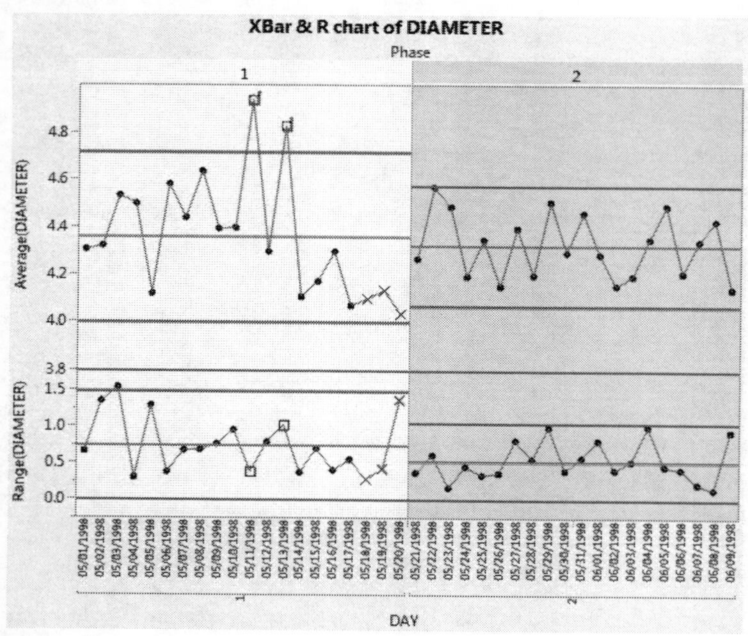

p-chart Example

The Washers.jmp sample data contains defect data for two different lot sizes from the *ASTM Manual on Presentation of Data and Control Chart Analysis*, American Society for Testing and

Materials. To view the differences between constant and variable sample sizes, you can compare charts for Lot Size and Lot Size 2.

1. Select **Help > Sample Data Library** and open Quality Control/Washers.jmp.
2. Select **Analyze > Quality and Process > Control Chart Builder.**
3. Drag # defective to the **Y** role.

   An Individual & Moving Range chart appears.
4. Select **Shewhart Attribute** from the drop down to change the chart to an attribute chart.

   A c-chart appears.
5. Change the **Sigma** to **Binomial** to change the chart to a np-chart.
6. Change the **Statistic** from **Count** to **Proportion** to change the chart to a p-chart.

Figure 3.11 p-chart of # defective

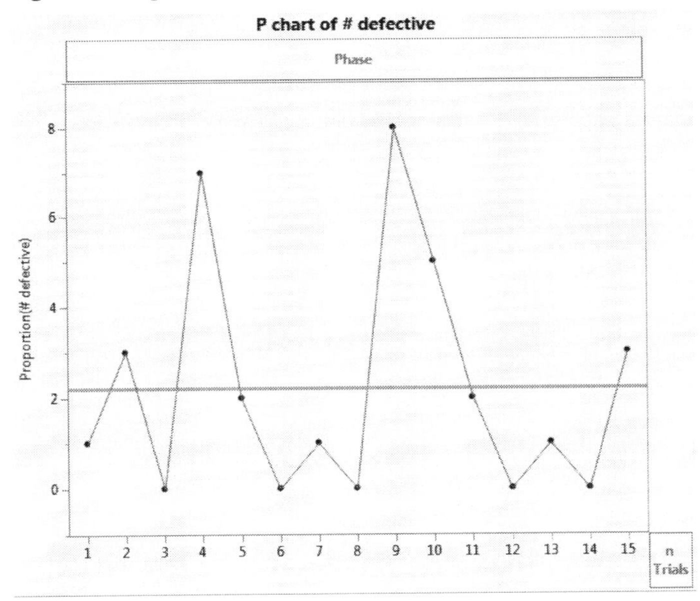

7. Drag Lot Size to the **nTrials** role.

Figure 3.12 p-chart of # defective with sample size

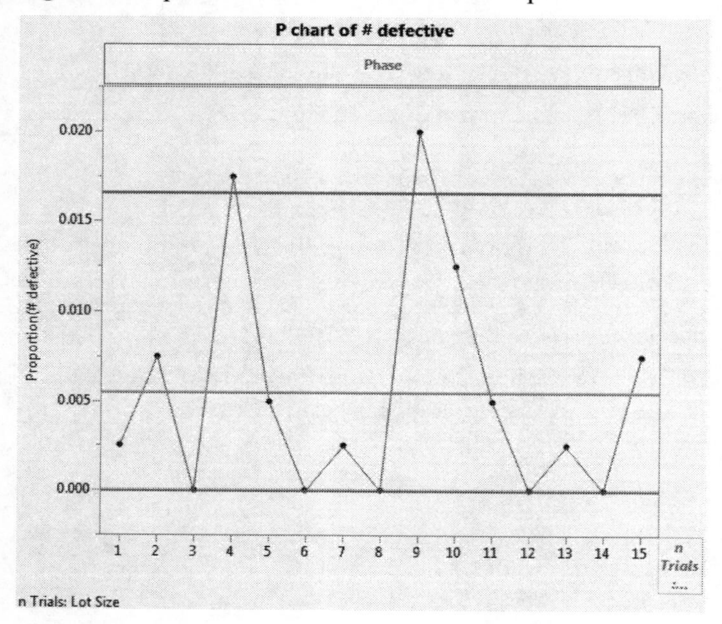

To view the differences between constant and variable sample sizes, you can compare charts for Lot Size and Lot Size 2 by simply dragging the variables to the nTrials zone.

np-chart Example

The Bottle Tops.jmp sample data contains simulated data from a bottle top manufacturing process. Sample is the sample ID number for each bottle. Status indicates whether the bottle top conformed to the design standards. In the Phase column, the first phase represents the time before the process adjustment. The second phase represents the time after the process adjustment. Notes on changes in the process are also included.

1. Select **Help > Sample Data Library** and open Quality Control/Bottle Tops.jmp.
2. Select **Analyze > Quality and Process > Control Chart Builder**.
3. Drag Sample to the **Subgroup** role.
4. Drag Status to the **Y** role.

Figure 3.13 np-chart of Status (Nonconforming)

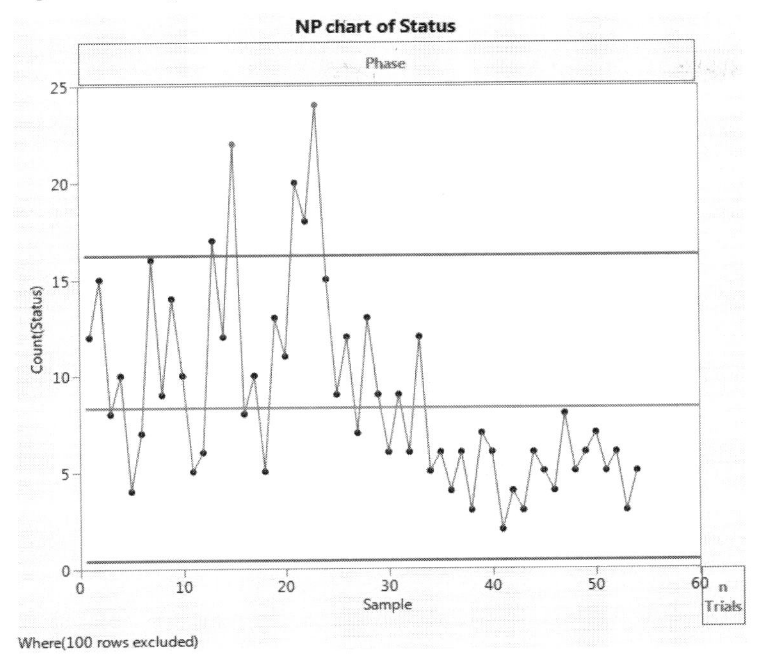

The original observations appear to have high variability and there are five observations (Samples 13, 15, 21, 22 and 23) that are outside of the upper control limit. Samples 15 and 23 note that new material and a new operator were introduced into the process, respectively. At the end of the phase, an adjustment was made to the manufacturing equipment. Therefore, the control limits for the entire series should not be used to assess the control during phase 2.

To compute separate control limits for each phase:

5. Drag Phase to the **Phase** zone.

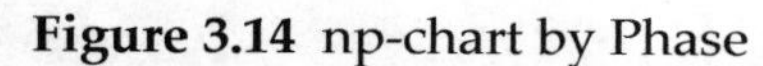

Figure 3.14 np-chart by Phase

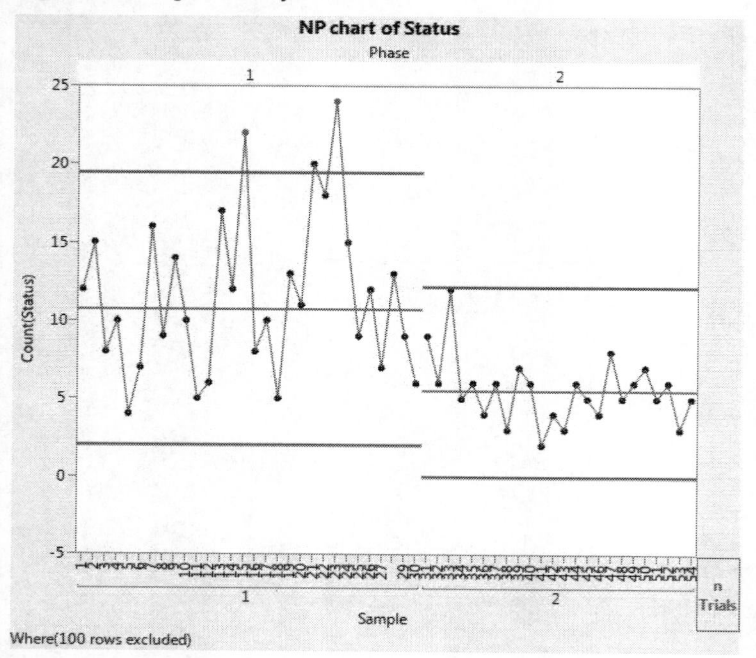

Including the **Phase** variable means that the control limits for phase 2 are based only on the data for phase 2. None of the phase 2 observations are outside the control limits. Therefore, you can conclude that the process is in control after the adjustment.

c-chart Example

The Cabinet Defects.jmp sample data table contains data concerning the various defects discovered while manufacturing cabinets over two time periods.

1. Select **Help > Sample Data Library** and open Quality Control/Cabinet Defects.jmp.
2. Select **Analyze > Quality and Process > Control Chart Builder.**
3. Drag Type of Defect to the **Y** role.
4. Drag Lot Number to the **Subgroup** role.

   A np-chart of Type of Defect appears.
5. To change to a c-chart, select **Poisson** from the **Sigma** list.
6. Open the **Type of Defect** disclosure button. Note all of the defect types are listed. Currently, only Bruised veneer is selected and displayed in the chart. You can select additional defect types and the chart updates immediately.

Figure 3.15 c-chart of Type of Defect

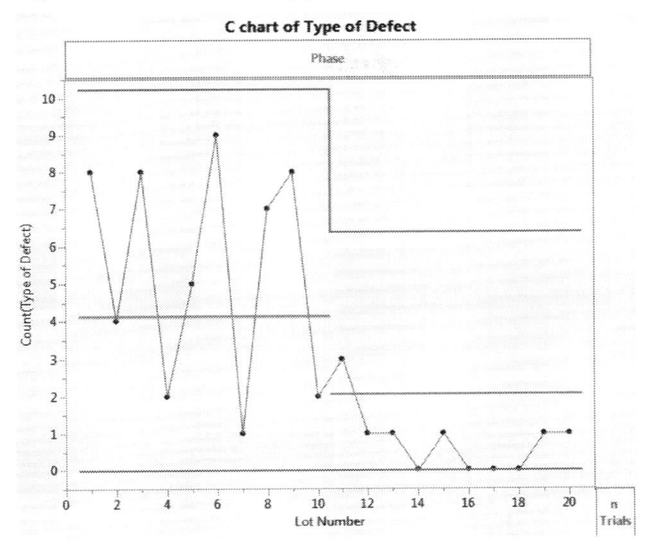

7. To add to phase variable, drag **Date** to the **Phase** zone.

Figure 3.16 c-chart of Type of Defect with Phases

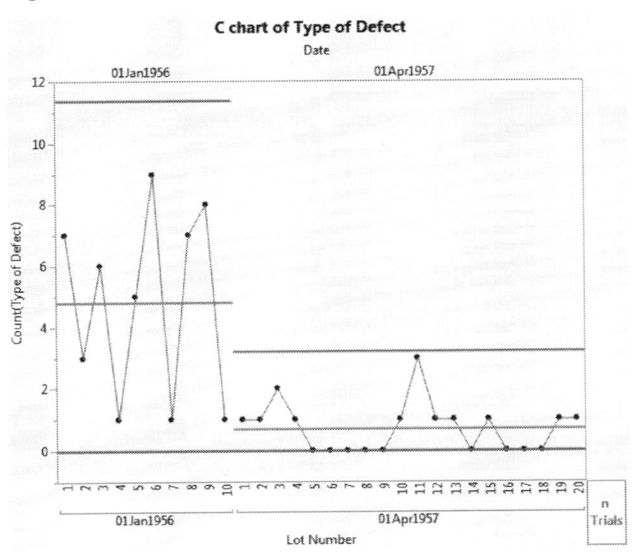

You can now view the results on the two different days. Both appear to be within limits. To examine other defect type behavior, select another defect type under the **Event Chooser** and view the results as the limits are updated.

u-chart Example

The Shirts.jmp sample data table contains data concerning the number of defects found in a number of boxes of shirts.

1. Select **Help > Sample Data Library** and open Quality Control/Shirts.jmp.
2. Select **Analyze > Quality and Process > Control Chart Builder.**
3. Drag # Defects to the **Y** role.
4. Drag Box to the **Subgroup** role.

   An Individual & Moving Range chart for # defects appears.
5. To change the chart to an Attribute chart, select **Shewhart Attribute** from the drop down list.

   A c-chart of # Defects appears.
6. Change the **Statistic** from **Count** to **Proportion** to change the chart to a u-chart.

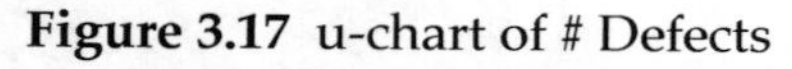

Figure 3.17 u-chart of # Defects

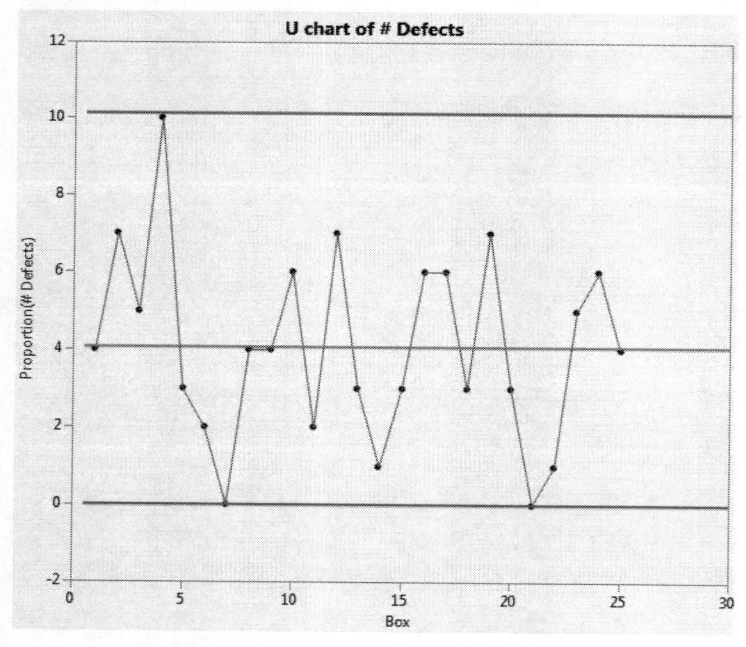

All of the points are within the control limits.

g-chart Example

Rare event charts are helpful when you know your data will not follow a normal distribution (for example, when measuring counts or wait times). The g-chart is an effective way to understand whether rare events are occurring more frequently than expected and warrant an intervention. A g-chart counts the number of possible opportunities since the last event. If you plot this type of data using a standard Shewhart control chart, you might see many more false signals, as the limits might be too narrow. The Adverse Reactions.jmp sample data table contains simulated data about adverse drug events (ADEs) reported by a group of hospital patients. An ADE is any type of injury or reaction the patient suffered after taking the drug. The date of the reaction and the number of days since the last reaction were recorded.

1. Select **Help > Sample Data Library** and open Quality Control/Adverse Reactions.jmp.
2. Select **Analyze > Quality and Process > Control Chart Builder**.
3. Drag Doses since Last ADE to the **Y** role.
4. Drag Date of ADE to the **Subgroup** role.

   An Individual & Moving Range chart of Doses since Last ADE appears.
5. To change the chart to a Rare Event chart, select **Rare Event** from the drop down list.

   A g-chart of Doses since Last ADE appears showing the number of doses given since the last event.

Figure 3.18 g-chart of Doses since Last ADE

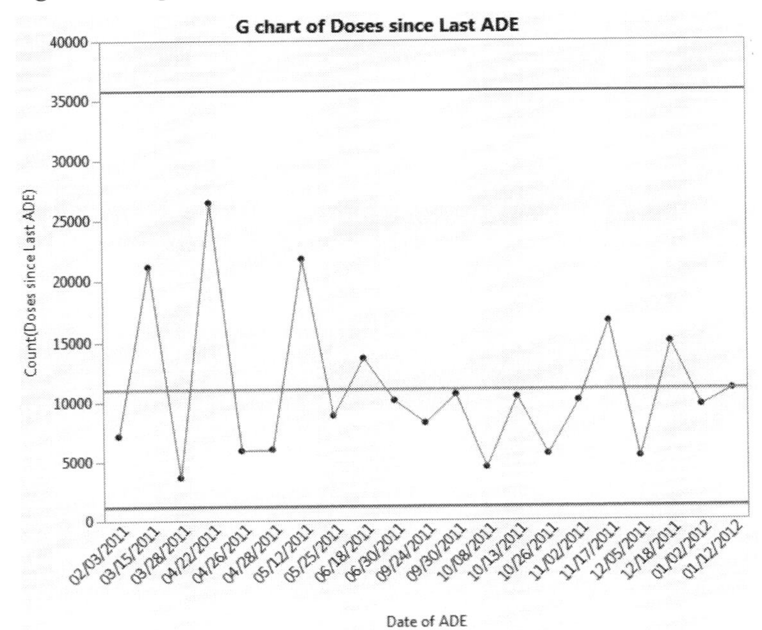

t-chart Example

Rare event charts are helpful when you know your data will not follow a normal distribution (for example, when measuring counts or wait times). t-charts are used to measure the time that has elapsed since the last event. If you plot this type of data using a standard Shewhart control chart, you might see many more false signals, as the limits might be too narrow. The Fan Burnout.jmp sample data table contains simulated data for a fan manufacturing process. The first column identifies each fan that burned out. The second column identifies the number of hours between each burnout.

1. Select **Help > Sample Data Library** and open Quality Control/Fan Burnout.jmp.
2. Select **Analyze > Quality and Process > Control Chart Builder.**
3. Drag Hours between Burnouts to the **Y** role.
4. Drag Burnout to the **Subgroup** role.

Figure 3.19 Individual and Moving Range Chart of Hours Between Burnouts

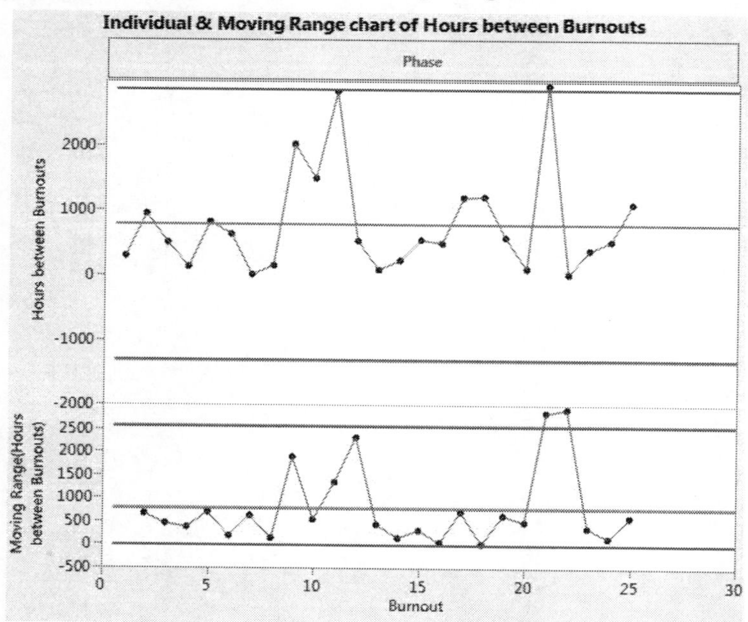

5. To change the chart to a Rare Event chart, select **Rare Event** from the drop down list.

   A g-chart of Hours between Burnouts appears. All points appear to be within the control limits.

6. Change the **Sigma** from **Negative Binomial** to **Weibull** to change the chart to a t-chart.

Figure 3.20 t-chart of Hours Between Burnouts

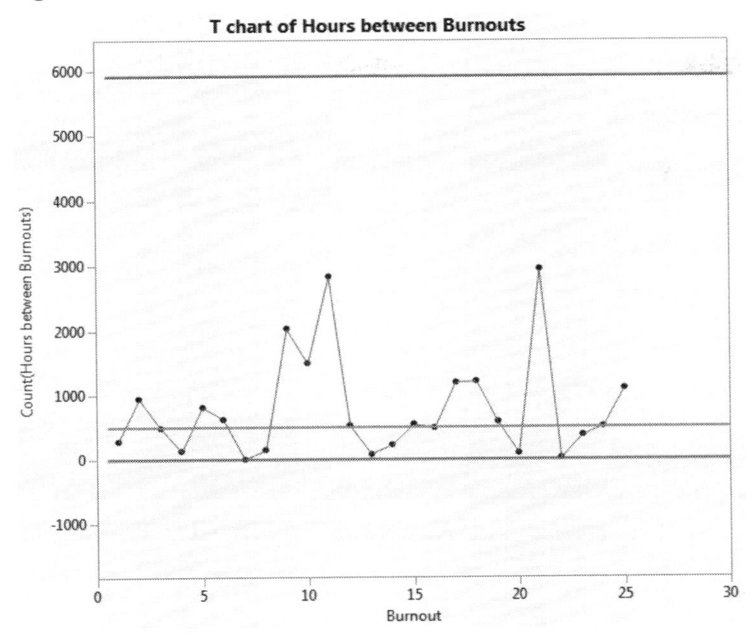

In the t-chart, all points appear to be within the control limits. It's clear that the Individual & Moving Range chart was inappropriate for the analysis, as the limits were too narrow.

Statistical Details for the Control Chart Builder Platform

This section contains statistical details for specific types of supported charts in the Control Chart Builder platform.

Control Limits for $\overline{X}$- and R-charts

JMP generates control limits for $\overline{X}$- and R-charts as:

$$\text{LCL for } \overline{X} \text{ chart} = \overline{X}_w - \frac{3\hat{\sigma}}{\sqrt{n_i}}$$

$$\text{UCL for } \overline{X} \text{ chart} = \overline{X}_w + \frac{3\hat{\sigma}}{\sqrt{n_i}}$$

LCL for *R*-chart $=\max\left(d_2(n_i)\hat{\sigma} - 3d_3(n_i)\hat{\sigma}, 0\right)$

UCL for *R*-chart $= d_2(n_i)\hat{\sigma} + 3d_3(n_i)\hat{\sigma}$

Center line for *R*-chart: By default, the center line for the ith subgroup (where 3 is the sigma multiplier) indicates an estimate of the expected value of R_i. This value is computed as:

$d_2(n_i)\hat{\sigma}$, where $\hat{\sigma}$ is an estimate of σ.

The standard deviation of an $\overline{X}/R$ chart is estimated by:

$$\hat{\sigma} = \frac{\frac{R_1}{d_2(n_1)} + \ldots + \frac{R_N}{d_2(n_N)}}{N}$$

where:

$\overline{X}_w$ = weighted average of subgroup means

σ = process standard deviation

n_i = sample size of ith subgroup

$d_2(n)$ is the expected value of the range of n independent normally distributed variables with unit standard deviation

$d_3(n)$ is the standard error of the range of n independent observations from a normal population with unit standard deviation

R_i is the range of ith subgroup

N is the number of subgroups for which $n_i \geq 2$

Control Limits for $\overline{X}$- and S-charts

JMP generates control limits for $\overline{X}$- and *S*-charts as:

LCL for $\overline{X}$ chart $= \overline{X}_w - \frac{3\hat{\hat{\sigma}}}{\sqrt{n_i}}$

UCL for $\overline{X}$ chart $= \overline{X}_w + \frac{3\hat{\hat{\sigma}}}{\sqrt{n_i}}$

$$\text{LCL for } S\text{-chart} = \max\left(c_4(n_i)\hat{\sigma} - 3c_5(n_i)\hat{\sigma}, 0\right)$$

$$\text{UCL for } S\text{-chart} = c_4(n_i)\hat{\sigma} + 3c_5(n_i)\hat{\sigma}$$

Center line for S-chart: By default, the center line for the ith subgroup (where 3 is the sigma multiplier) indicates an estimate of the expected value of s_i. This value is computed as $c_4(n_i)\hat{\sigma}$, where $\hat{\sigma}$ is an estimate of σ.

The estimate for the standard deviation in an $\bar{X}/S$ chart is:

$$\hat{\sigma} = \frac{\dfrac{s_1}{c_4(n_1)} + \ldots + \dfrac{s_N}{c_4(n_N)}}{N}$$

where:

$\bar{X}_w$ = weighted average of subgroup means

σ = process standard deviation

n_i = sample size of ith subgroup

$c_4(n)$ is the expected value of the standard deviation of n independent normally distributed variables with unit standard deviation

$c_5(n)$ is the standard error of the standard deviation of n independent observations from a normal population with unit standard deviation

N is the number of subgroups for which $n_i \geq 2$

s_i is the sample standard deviation of the ith subgroup

Control Limits for Individual Measurement and Moving Range Charts

$$\text{LCL for Individual Measurement Chart} = \bar{X} - 3\hat{\sigma}$$

$$\text{UCL for Individual Measurement Chart} = \bar{X} + 3\hat{\sigma}$$

$$\text{LCL for Moving Range Chart} = \max\left(d_2(2)\hat{\sigma} - 3d_3(2)\hat{\sigma}, 0\right)$$

$$\text{UCL for Moving Range Chart} = d_2(2)\hat{\sigma} + 3d_3(2)\hat{\sigma}$$

The standard deviation for Individual Measurement and Moving Range charts is estimated by:

$$\hat{\sigma} = \frac{\overline{MR}}{d_2(2)}$$

where:

$\overline{X}$ = the mean of the individual measurements

$\overline{MR}$ = the mean of the nonmissing moving ranges computed as $(MR_2+MR_3+...+MR_N)/(N-1)$ where $MR_i = |x_i - x_{i-1}|$.

σ = the process standard deviation

$d_2(2)$ = expected value of the range of two independent normally distributed variables with unit standard deviation.

$d_3(2)$ = standard error of the range of two independent observations from a normal population with unit standard deviation.

Control Limits for p- and np-charts

The lower and upper control limits, LCL, and UCL, respectively, are computed as:

p-chart LCL $= \max(\bar{p} - 3\sqrt{\bar{p}(1-\bar{p})/n_i}, 0)$

p-chart UCL $= \min(\bar{p} + 3\sqrt{\bar{p}(1-\bar{p})/n_i}, 1)$

np-chart LCL $= \max(n_i\bar{p} - 3\sqrt{n_i\bar{p}(1-\bar{p})}, 0)$

np-chart UCL $= \min(n_i\bar{p} + 3\sqrt{n_i\bar{p}(1-\bar{p})}, n_i)$

where:

$\bar{p}$ is the average proportion of nonconforming items taken across subgroups

$$\bar{p} = \frac{n_1p_1 + \ldots + n_Np_N}{n_1 + \ldots + n_n} = \frac{X_1 + \ldots + X_N}{n_1 + \ldots + n_N}$$

n_i is the number of items in the ith subgroup

3 is the number of standard deviations

Control Limits for u-charts

The lower and upper control limits, LCL, and UCL, are computed as:

$$\text{LCL} = \max(\bar{u} - 3\sqrt{\bar{u}/n_i}, 0)$$

$$\text{UCL} = \bar{u} + 3\sqrt{\bar{u}/n_i}$$

The limits vary with n_i.

u is the expected number of nonconformities per unit produced by process

u_i is the number of nonconformities per unit in the ith subgroup. In general, $u_i = c_i/n_i$.

c_i is the total number of nonconformities in the ith subgroup

n_i is the number of inspection units in the ith subgroup

$\bar{u}$ is the average number of nonconformities per unit taken across subgroups. The quantity $\bar{u}$ is computed as a weighted average

$$\bar{u} = \frac{n_1 u_1 + \ldots + n_N u_N}{n_1 + \ldots + n_N} = \frac{c_1 + \ldots + c_N}{n_1 + \ldots + n_N}$$

N is the number of subgroups

Control Limits for c-charts

The lower and upper control limits, LCL, and UCL, are computed as:

$$\text{LCL} = \max(n_i\bar{u} - 3\sqrt{n_i\bar{u}}, 0)$$

$$\text{UCL} = n_i\bar{u} + 3\sqrt{n_i\bar{u}}$$

The limits vary with n_i.

u is the expected number of nonconformities per unit produced by process

u_i is the number of nonconformities per unit in the ith subgroup. In general, $u_i = c_i/n_i$.

c_i is the total number of nonconformities in the ith subgroup

n_i is the number of inspection units in the ith subgroup

$\bar{u}$ is the average number of nonconformities per unit taken across subgroups. The quantity $\bar{u}$ is computed as a weighted average

$$\bar{u} = \frac{n_1 u_1 + \ldots + n_N u_N}{n_1 + \ldots + n_N} = \frac{c_1 + \ldots + c_N}{n_1 + \ldots + n_N}$$

N is the number of subgroups

Levey-Jennings Charts

Levey-Jennings charts show a process mean with control limits based on a long-term sigma. The control limits are placed at 3*s* distance from the center line.

The standard deviation, *s,* for the Levey-Jennings chart is calculated the same way standard deviation is in the Distribution platform.

$$s = \sqrt{\sum_{i=1}^{N} \frac{(y_i - \bar{y})^2}{N-1}}$$

Control Limits for g-charts

The negative binomial distribution is an extension of the geometric (Poisson) distribution and allows for over-dispersion relative to the Poisson. The negative binomial distribution can be used to construct both exact and approximate control limits for count data. Approximate control limits can be obtained based on a chi-square approximation to the negative binomial. All data is used as individual observations regardless of subgroup size.

Let X have a negative binomial distribution with parameters (u, 3). Then:

$$P(X \le r) \sim P\left(X^2_v \le \frac{2r+1}{1+3u}\right)$$

where:

X^2_v is a chi-square variate with $v = 2u/(1+3u)$ degrees of freedom.

Based on this approximation, approximate upper and lower control limits can be determined. For a nominal level α Type 1 error probability in one direction, an approximate upper control limit is a limit UCL such that:

$$P(X > \text{UCL}) = 1 - P\left(X^2_v \le \frac{2UCL+1}{1+3u}\right) = \alpha$$

Likewise, an approximate lower control limit, LCL, is a limit such that:

$$P(X < \text{LCL}) = 1 - P\left(X^2_v \ge \frac{2LCL+1}{1+3u}\right) = \alpha$$

Thus, an approximate level lower and upper control limits, LCL and UCL, respectively, are computed as:

$$\text{UCL} = \frac{X^2_{v,1-\alpha}(1+3u)-1}{2}$$

$$\text{LCL} = \frac{X^2_{v,\alpha}(1+3u)-1}{2}$$

where:

$X^2_{v, 1-\alpha}(X^2_{v, \alpha})$ is the upper (lower) percentile of the chi-square distribution with $v = 2u/(1+3u)$ degrees of freedom. Negative lower control limits can be set to zero.

Control Limits for t-charts

If there are no 0's in the data, the estimates of the shape and scale parameters are calculated from the data and used to obtain the percentiles of the Weibull distribution.

To estimate limits from the data:

If

p1 = normalDist(-3) for Normal (0,1)

p2 = normalDist(0) for Normal (0,1)

p3 = normalDist(3) for Normal (0,1)

Then

CL = Weibull Quantile (p2, β) * α

UCL = Weibull Quantile (p1, β) * α

LCL = Weibull Quantile (p3, β) * α

where:

β is the shape parameter and α is the scale parameter for the Weibull Quantile function. For more information about the Weibull Quantile function, see Help > Scripting Index.

Chapter 4

Shewhart Control Charts

Create Variable and Attribute Control Charts

A control chart is a graphical and analytic tool for monitoring process variation. The natural variation in a process can be quantified using a set of control limits. Control limits help distinguish common-cause variation from special-cause variation. Typically, action is taken to eliminate special-cause variation and bring the process back in control. It is also important to quantify the common-cause variation in a process, as this determines process capability.

The Control Chart platform in JMP provides a variety of control charts, as well as run charts. To support process improvement initiatives, most of the control chart options display separate control charts for different phases of a project on the same chart.

- Run Chart
- $\overline{X}$-, *R*-, and *S*-charts
- Individual and Moving Range Charts
- *p*-, *np*-, *c*-, and *u*-charts
- UWMA and EWMA Charts
- CUSUM Charts
- Presummarize, Levey-Jennings, and Multivariate Control Charts
- Phase Control Charts for $\overline{X}$-, *R*-, *S*-, *IR*-, *p*-, *np*-, *c*-, *u*-, Presummarize, and Levey-Jennings Charts.

Figure 4.1 Control Chart Example

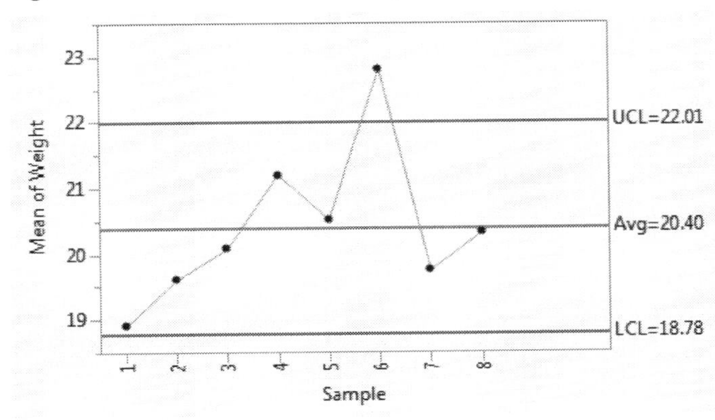

Overview of the Control Chart Platform

A control chart is a graphical way to filter out routine variation in a process. Filtering out routine variation helps manufacturers and other businesses determine whether a process is stable and predictable. If the variation is more than routine, the process can be adjusted to create higher quality output at a lower cost.

All processes exhibit variation as the process is measured over time. There are two types of variation in process measurements:

- *Routine* or *common-cause* variation. Even measurements from a stable process exhibit these random ups and downs. When process measurements exhibit only common-cause variation, the measurements stay within acceptable limits.
- *Abnormal* or *special-cause* variation. Examples of special-cause variation include a change in the process mean, points above or below the control limits, or measurements that trend up or down. These changes can be caused by factors such as a broken tool or machine, equipment degradation, and changes to raw materials. A change or defect in the process is often identifiable by abnormal variation in the process measurements.

Control charts quantify the routine variation in a process, so that special causes can be identified. One way control charts filter out routine variation is by applying control limits. Control limits define the range of process measurements for a process that is exhibiting only routine variation. Measurements between the control limits indicate a stable and predictable process. Measurements outside the limits indicate a special cause, and action should be taken to restore the process to a state of control.

Control chart performance is dependent on the sampling scheme used. The sampling plan should be *rational,* that is, the subgroups are representative of the process. *Rational subgrouping* means that you sample from the process by selecting subgroups in such a way that special causes are more likely to occur between subgroups rather than within subgroups.

Shewhart control charts are broadly classified into control charts for variables and control charts for attributes. Control charts for variables include moving average and CUSUM charts. CUSUM charts are also a type of attribute chart. For details, see "Moving Average Charts" on page 79 and the "Cumulative Sum Control Charts" chapter on page 113.

Example of the Control Chart Platform

The following example uses the Coating.jmp sample data table in the Quality Control sample data folder (taken from the *ASTM Manual on Presentation of Data and Control Chart Analysis*). The quality characteristic of interest is the Weight column. A subgroup sample of four is chosen.

1. Select **Help > Sample Data Library** and open Quality Control/Coating.jmp.

2. Select **Analyze > Quality And Process > Control Chart > XBar.**

   Note the selected chart types of **XBar** and **R.**

3. Select Weight and click **Process.**
4. Select Sample and click **Sample Label.**
5. Click **OK.**

Figure 4.2 Variables Charts for Coating Data

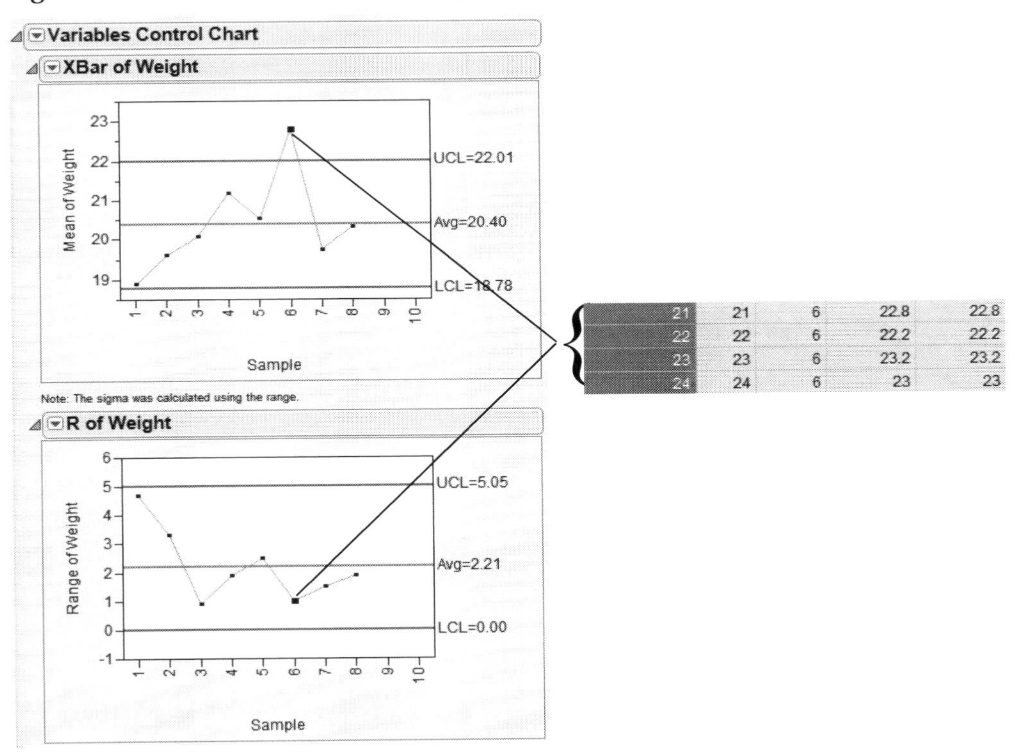

An $\overline{X}$-chart and an R-chart for the process are shown in Figure 4.2. Sample six indicates that the process is not in statistical control. To check the sample values, click the sample six summary point on either control chart. The corresponding rows highlight in the data table.

Note: If an S chart is chosen with the $\overline{X}$-chart, then the limits for the $\overline{X}$-chart are based on the standard deviation. Otherwise, the limits for the $\overline{X}$-chart are based on the range.

Shewhart Control Chart Types

Shewhart control charts are broadly classified into control charts for variables and control charts for attributes.

Control Charts for Variables

Control charts for variables are classified according to the subgroup summary statistic plotted on the chart:

- Run charts display data as a connected series of points
- $\overline{X}$-charts display subgroup means (averages)
- *R*-charts display subgroup ranges (maximum – minimum)
- *S*-charts display subgroup standard deviations
- Presummarize charts display subgroup means and standard deviations

The **IR** selection gives additional chart types:

- Individual Measurement charts display individual measurements
- Moving Range charts display moving ranges of two or more successive measurements

Run Charts

Run charts display a column of data as a connected series of points. Run charts can also plot the group means when the **Sample Label** role is used, either on the window or through a script.

XBar-, R-, and S- Charts

For quality characteristics measured on a continuous scale, a typical analysis shows both the process mean and its variability with a mean chart aligned above its corresponding *R*- or *S*-chart.

Individual Measurement Charts

Individual Measurement charts displays individual measurements. Individual Measurement charts are appropriate when only one measurement is available for each subgroup sample.

Moving Range charts displays moving ranges of two or more successive measurements. Moving ranges are computed for the number of consecutive measurements that you enter in the **Range Span** box. The default range span is 2. Because moving ranges are correlated, these charts should be interpreted with care.

Moving Average Charts

The control charts previously discussed plot each point based on information from a single subgroup sample. The Moving Average chart is different from other types because each point combines information from the current sample and from past samples. As a result, the Moving Average chart is more sensitive to small shifts in the process average. On the other hand, it is more difficult to interpret patterns of points on a Moving Average chart because consecutive moving averages can be highly correlated (Nelson 1982).

In a Moving Average chart, the quantities that are averaged can be individual observations instead of subgroup means. However, a Moving Average chart for individual measurements is not the same as a control (Shewhart) chart for individual measurements or moving ranges with individual measurements plotted.

Uniformly Weighted Moving Average Charts

Each point on a Uniformly Weighted Moving Average (UWMA) chart, also called a Moving Average chart, is the average of the w most recent subgroup means, including the present subgroup mean. When you obtain a new subgroup sample, the next moving average is computed by dropping the oldest of the previous w subgroup means and including the newest subgroup mean. The constant, w, is called the *span* of the moving average, and indicates how many subgroups to include to form the moving average. The larger the span (w), the smoother the UWMA line, and the less it reflects the magnitude of shifts. This means that larger values of w guard against smaller shifts.

Exponentially Weighted Moving Average Charts

Each point on an Exponentially Weighted Moving Average (EWMA) chart, also referred to as a Geometric Moving Average (GMA) chart, is the weighted average of all the previous subgroup means, including the mean of the present subgroup sample. The weights decrease exponentially going backward in time. The weight ($0 < \text{weight} \leq 1$) assigned to the present subgroup sample mean is a parameter of the EWMA chart. Small values of weight are used to guard against small shifts.

Presummarize Charts

If your data consist of repeated measurements of the same process unit, you can combine these into one measurement for the unit. Pre-summarizing is not recommended unless the data have repeated measurements on each process or measurement unit.

Presummarize summarizes the process column into sample means and/or standard deviations, based either on the sample size or sample label chosen. Then it charts the summarized data based on the options chosen in the launch window. You can also append a capability analysis by checking the appropriate box in the launch window.

Control Charts for Attributes

In the previous types of charts, measurement data was the process variable. This data is often continuous, and the charts are based on theory for continuous data. Another type of data is count data, where the variable of interest is a discrete count of the number of defects or blemishes per subgroup. For discrete count data, attribute charts are applicable, as they are based on binomial and Poisson models. Because the counts are measured per subgroup, it is important when comparing charts to determine whether you have a similar number of items in the subgroups between the charts. Attribute charts, like variables charts, are classified according to the subgroup sample statistic plotted on the chart.

Determining Which Attribute Chart to Use

Each item is judged as either conforming or non-conforming:

p-chart Shows the *proportion* of defective items.

np-chart Shows the *number* of defective items.

The number of defects is counted for each item:

c-chart Shows the *number* of defective items.

u-chart Shows the *average number* of defective items.

For attribute charts, specify the column containing the defect count or defective proportion as the Process variable. The data are interpreted as counts, unless the column contains non-integer values between 0 and 1.

- p-charts display the proportion of nonconforming (defective) items in subgroup samples, which can vary in size. Since each subgroup for a p-chart consists of N_i items, and an item is judged as either conforming or nonconforming, the maximum number of nonconforming items in a subgroup is N_i.
- np-charts display the number of nonconforming (defective) items in subgroup samples. Because each subgroup for a np-chart consists of N_i items, and an item is judged as either conforming or nonconforming, the maximum number of nonconforming items in subgroup i is N_i.

Note: To use the Sigma column property for *P-* or *NP-* charts, the value needs to be equal to the proportion. JMP calculates the sigma as a function of the proportion and the sample sizes.

- c-charts display the number of nonconformities (defects) in a subgroup sample that usually, but does not necessarily, consists of one inspection unit.

Caution: For a c-chart, if you do not specify a Sample Size or Constant Size, then the Sample Label is used as the sample size.

- u-charts display the number of nonconformities (defects) per unit in subgroup samples that can have a varying number of inspection units.

Caution: For a u-chart, if you do not specify a Unit Size or Constant Size, then the Sample Label is used as the unit size.

Levey-Jennings Charts

Levey-Jennings charts show a process mean with control limits based on a long-term sigma. The control limits are placed at 3*s* distance from the center line. The standard deviation, *s*, for the Levey-Jennings chart is calculated the same way standard deviation is in the Distribution platform.

Launch the Control Chart Platform

When you launch the Control Chart platform by selecting **Analyze > Quality And Process > Control Chart**, you will see a Control Chart Launch window similar to Figure 4.3. The exact controls vary depending on which type of chart you select. Initially, the window shows the following types of information:

- Process information, for measurement variable selection
- Chart type information
- Limits specifications
- Specified statistics

Specific information shown for each section varies according to the type of chart that you select. Through interaction with the Launch window, you specify exactly how you want your charts to be created. The following sections describe the window elements.

Figure 4.3 XBar Control Chart Launch Window

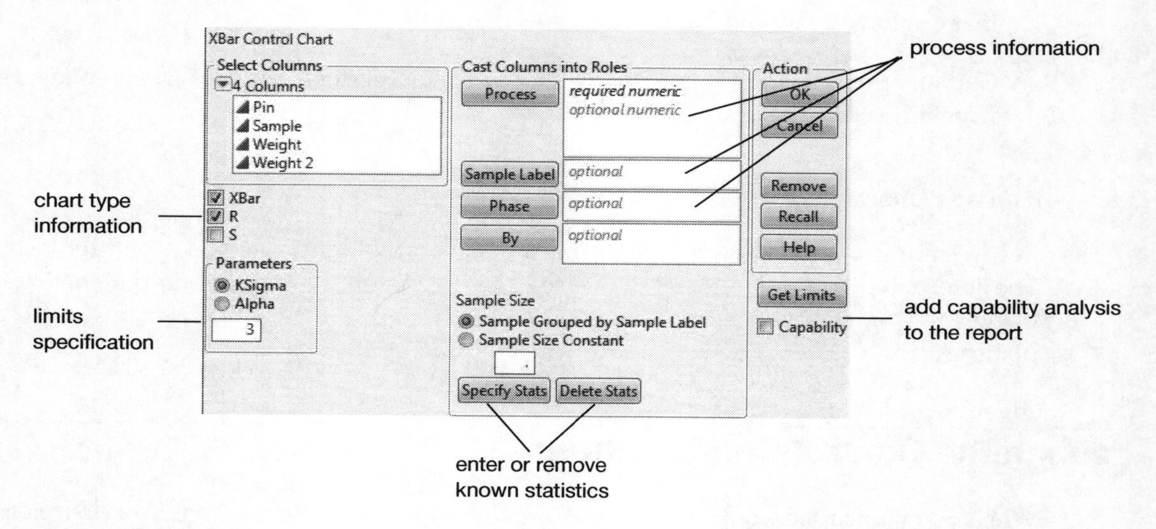

Process Information

The Launch window displays a list of columns in the current data table. Here, you specify the variables to be analyzed and the subgroup sample size.

Process

The **Process** role selects variables for charting.

- For variables charts, specify measurements as the process.
- For attribute charts, specify the defect count or defective proportion as the process. The data are interpreted as counts, unless it contains non-integer values between 0 and 1.

Note: The rows of the table must be sorted in the order in which you want them to appear in the control chart. Even if there is a **Sample Label** variable specified, you still must sort the data accordingly.

Sample Label

The **Sample Label** role enables you to specify a variable whose values label the horizontal axis and can also identify unequal subgroup sizes. If no sample label variable is specified, the samples are identified by their subgroup sample number.

- If the sample subgroups are the same size, select the **Sample Size Constant** option and enter the size in the text box. If you entered a Sample Label variable, its values are used to label the horizontal axis. The sample size is used in the calculation of the limits regardless of whether the samples have missing values.

- If the sample subgroups have an unequal number of rows or have missing values and you have a column identifying each sample, select the **Sample Grouped by Sample Label** option and enter the sample identifying column as the sample label.

For attribute charts (p-, np-, c-, and u-charts), this variable is the subgroup sample size. Additional options appear on the launch window, including **Sample Size**, **Constant Size**, and/or **Unit Size**, depending on your selection. In variables charts, it identifies the sample. When the chart type is **IR**, a **Range Span** text box appears. The *range span* specifies the number of consecutive measurements from which the moving ranges are computed.

Note: The rows of the table must be sorted in the order in which you want them to appear in the control chart. Even if there is a **Sample Label** variable specified, you still must sort the data accordingly.

The illustration in Figure 4.4 shows an $\overline{X}$-chart for a process with unequal subgroup sample sizes, using the Coating.jmp sample data from the Quality Control sample data folder.

Figure 4.4 Variables Charts with Unequal Subgroup Sample Sizes

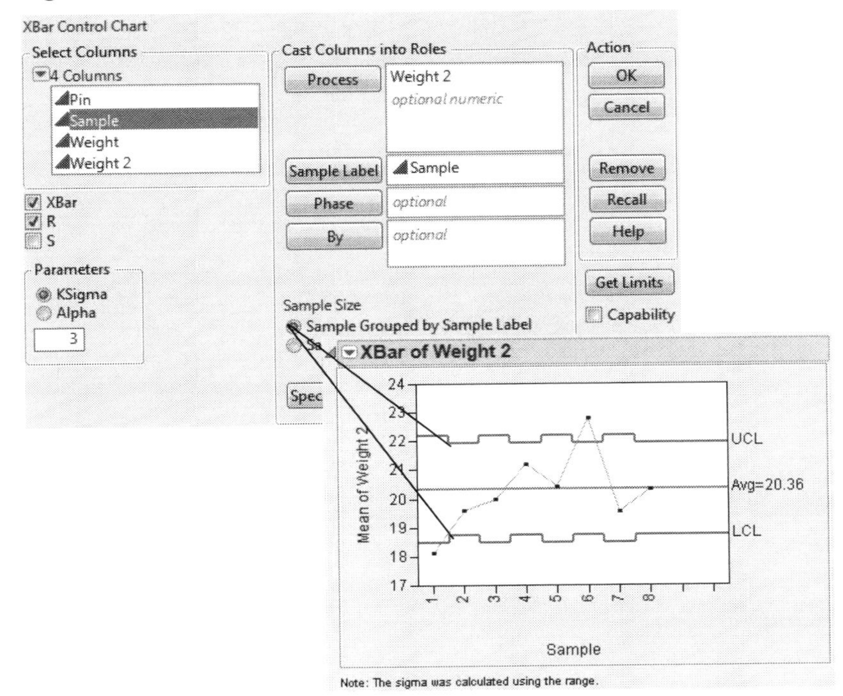

Phase

The **Phase** role enables you to specify a column identifying different phases, or sections. A *phase* is a group of consecutive observations in the data table. For example, phases might correspond to time periods during which a new process is brought into production and then put through successive changes. Phases generate, for each level of the specified Phase variable, a new sigma, set of limits, zones, and resulting tests.

On the window for $\overline{X}$-, *R*-, S-, *IR*-, *P*-, *NP*-, *C*-, *U*-, Presummarize, and Levey-Jennings charts, a **Phase** variable button appears. If a phase variable is specified, the phase variable is examined, row by row, to identify to which phase each row belongs. Saving to a limits file reveals the sigma and specific limits calculated for each phase. See "Phase Example" on page 109 for an example.

By

The **By** role identifies a variable to produce a separate analysis for each value that appears in the column.

Chart Type Information

Shewhart control charts are broadly classified as variables charts and attribute charts. Moving average charts and CUSUM charts can be thought of as special types of variables charts.

Figure 4.5 Window Options for Variables Control Charts

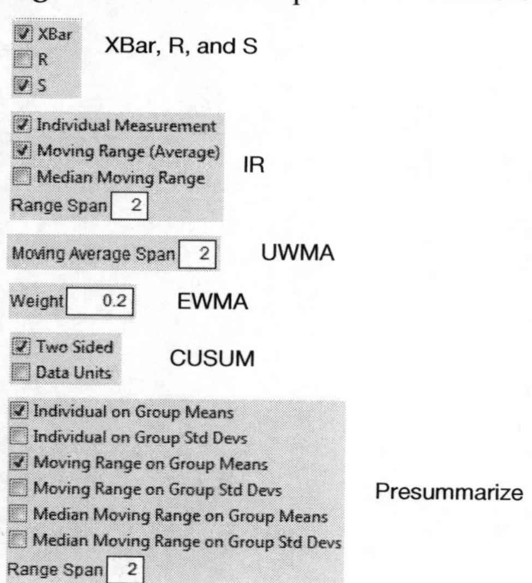

- **XBar** charts menu selection gives **XBar**, **R**, and **S** check boxes.

- The **IR** menu selection has check box options for the Individual Measurement, Moving Range, and Median Moving Range charts.
- The uniformly weighted moving average (**UWMA**) and exponentially weighted moving average (**EWMA**) selections are special charts for means.
- The **CUSUM** chart is a special chart for means or individual measurements.
- **Presummarize** enables you to specify information about pre-summarized statistics.
- **P**, **NP**, **C**, and **U** charts, **Run Chart**, and **Levey-Jennings** charts have no additional specifications.

The types of control charts are discussed in "Overview of the Control Chart Platform" on page 76.

Limits Specifications

You can specify computations for control limits by entering a value for *k* (**K Sigma**), or by entering a probability for α(**Alpha**), or by retrieving a limits value from the process columns' properties or a previously created Limits Table. Limits Tables and the **Get Limits** button are discussed in the section "Saving and Retrieving Limits" on page 93. There must be a specification of either **K Sigma** or **Alpha**. The window default for **K Sigma** is 3.

KSigma

The **KSigma** parameter option allows specification of control limits in terms of a multiple of the sample standard error. **KSigma** specifies control limits at *k* sample standard errors above and below the expected value, which shows as the center line. To specify *k*, the number of sigmas, click the radio button for **KSigma** and enter a positive *k* value into the text box. The usual choice for *k* is 3, which is three standard deviations. The examples shown in Figure 4.6 compare the $\overline{X}$-chart for the Coating.jmp data with control lines drawn with **KSigma** = 3 and **KSigma** = 4.

Figure 4.6 K Sigma =3 (left) and K Sigma=4 (right) Control Limits

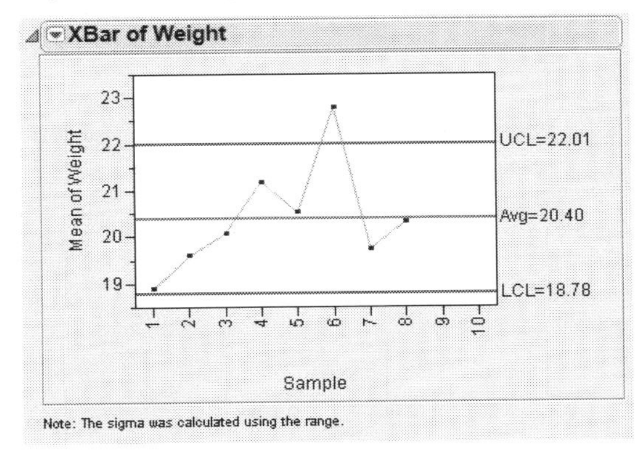

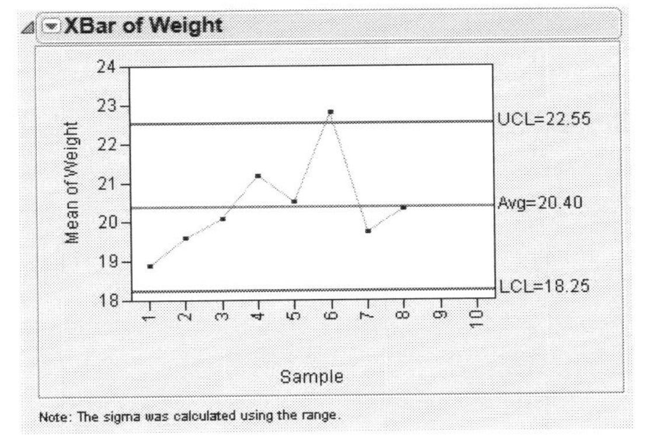

Alpha

The **Alpha** parameter option specifies control limits (also called *probability limits*) in terms of the probability α that a single subgroup statistic exceeds its control limits, assuming that the process is in control. To specify alpha, click the **Alpha** radio button and enter the probability that you want. Reasonable choices for α are 0.01 or 0.001. The **Alpha** value equivalent to a **KSigma** of 3 is 0.0027.

Specified Statistics

After specifying a process variable, if you click the **Specify Stats** (when available) button on the Control Chart Launch window, a tab with editable fields is appended to the bottom of the window. This lets you enter historical statistics (that is, statistics obtained from historical data) for the process variable. The Control Chart platform uses those entries to construct control charts. The example here shows 1 as the standard deviation of the process variable and 20 as the mean measurement.

Figure 4.7 Example of Specify Stats

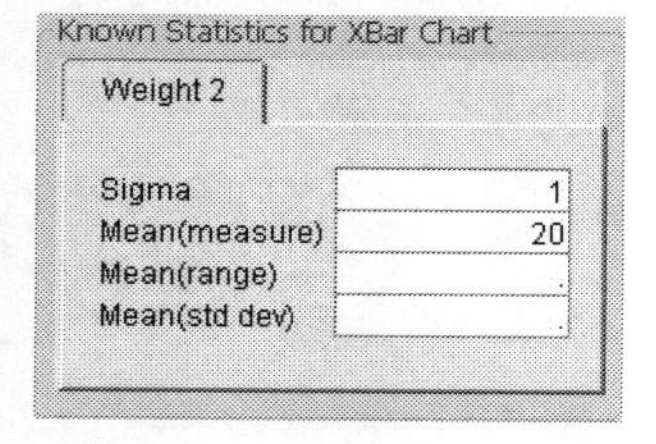

Note: When the mean is user-specified, it is labeled in the plot as μ0.

If you check the **Capability** option on the Control Chart launch window (see Figure 4.3), a window appears as the platform is launched asking for specification limits. The standard deviation for the control chart selected is sent to the window and appears as a Specified Sigma value, which is the default option. After entering the specification limits and clicking OK, capability output appears in the same window next to the control chart. For information about how the capability indices are computed, see the Distributions chapter in the *Basic Analysis* book.

The Control Chart Report

The analysis produces a chart that can be used to determine whether a process is in a state of statistical control. The report varies depending on the type of chart that you select. Figure 4.8 displays the parts of a simple control chart. Control charts update dynamically as data is added or changed in the data table.

Figure 4.8 Example of a Control Chart

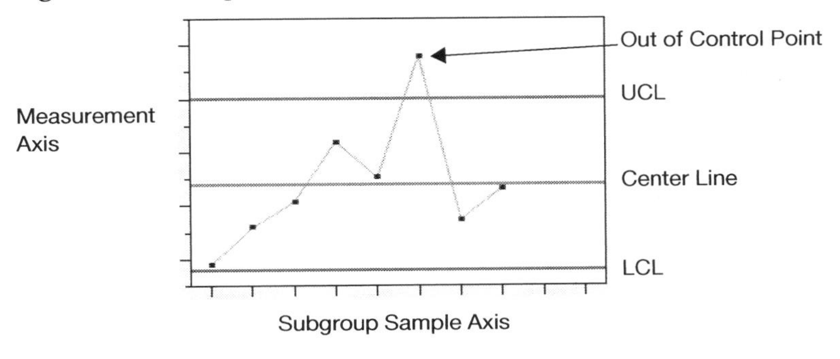

Note: Any rows that are excluded in the data table are also hidden in Run charts, P-charts, U-charts, and C-charts.

Control charts have the following characteristics:

- Each point plotted on the chart represents an individual process measurement or summary statistic. In Figure 4.8, the points represent the average for a sample of measurements.

  Subgroups should be chosen *rationally*, that is, they should be chosen to maximize the probability of seeing a true process signal *between* subgroups.
- The vertical axis of a control chart is scaled in the same units as the summary statistic.
- The horizontal axis of a control chart identifies the subgroup samples and is time ordered. Observing the process over time is important in assessing if the process is changing.
- The green line is the center line, or the average of the data. The center line indicates the average (expected) value of the summary statistic when the process is in statistical control. Measurements should appear equally on both sides of the center line. If not, this is possible evidence that the process average is changing.
- The two red lines are the upper and lower control limits, labeled UCL and LCL. These limits give the range of variation to be expected in the summary statistic when the process is in statistical control. If the process is exhibiting only routine variation, then all the points should fall randomly in that range. In Figure 4.8, one measurement is above the upper control limit. This is evidence that the measurement could have been influenced by a special cause, or is possibly a defect.
- A point outside the control limits (or the V-mask of a CUSUM chart) signals the presence of a special cause of variation.

Options within each platform create control charts that can be updated dynamically as samples are received and recorded or added to the data table.

When a control chart signals abnormal variation, action should be taken to return the process to a state of statistical control if the process degraded. If the abnormal variation indicates an improvement in the process, the causes of the variation should be studied and implemented.

When you double-click the *x* or *y* axis, the appropriate Axis Specification window appears for you to specify the format, axis values, number of ticks, gridline, reference lines, and other options to display on the axis.

For example, the Pickles.jmp data lists measurements taken each day for three days. In Figure 4.9, by default, the *x* axis is labeled at every other tick. Sometimes this gives redundant labels, as shown to the left in Figure 4.9. If you specify a label at an increment of eight, the *x* axis is labeled once for each day, as shown in the chart on the right.

Figure 4.9 Example of Labeled x Axis Tick Marks

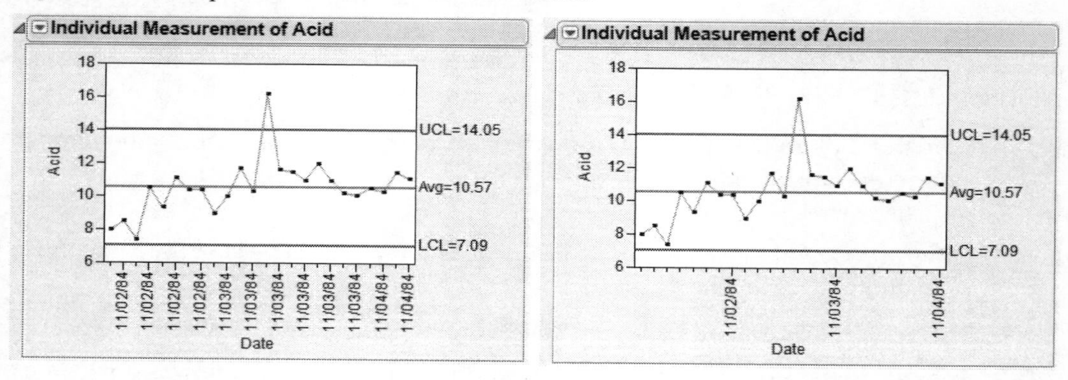

Tip: For information about warnings and rules, see "Tests" on page 48 and "Westgard Rules" on page 51 in the "Control Chart Builder" chapter of this guide.

Control Chart Platform Options

Control Charts have red triangle menus that affect various parts of the platform:

- The menu on the top-most title bar affects the whole platform window. Its items vary with the type of chart that you select.
- There is a menu of items on the chart type title bar with options that affect each chart individually.

Control Chart Window Options

The red triangle menu on the window title bar lists options that affect the report window. If you request **XBar** and **R** at the same time, you can check each chart type to show or hide it. The specific options that are available depend on the type of control chart you request. Unavailable options show as grayed menu items.

Show Limits Legend Shows or hides the Avg, UCL, and LCL values to the right of the chart.

Connect Through Missing Connects points when some samples have missing values. In Figure 4.10, the left chart has no missing points. The middle chart has samples 2, 11, 19, and 27 missing with the points not connected. The right chart appears if you select the **Connect Through Missing** option, which is the default.

Figure 4.10 Example of Connected through Missing Option

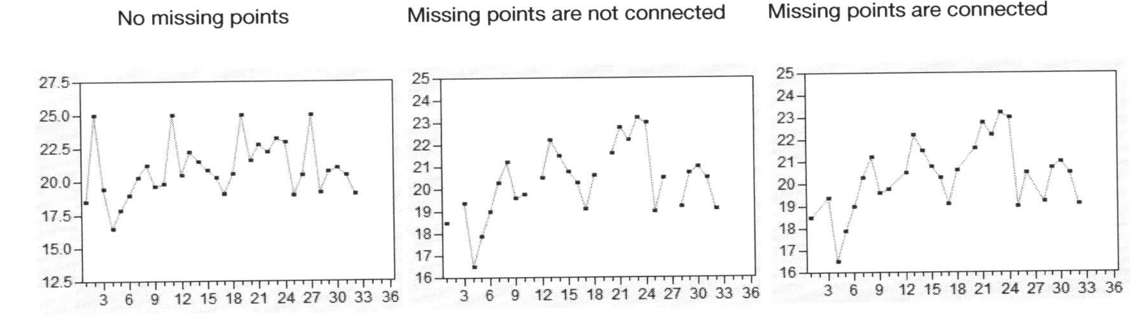

Use Median For Run Charts, when you select the **Show Center Line** option in the individual Run Chart red triangle menu, a line is drawn through the center value of the column. The center line is determined by the **Use Median** setting of the main Run Chart red triangle menu. When **Use Median** is selected, the median is used as the center line. Otherwise, the mean is used. When saving limits to a file, both the overall mean and median are saved.

Capability Performs a Capability Analysis for your data. A popup window is first shown, where you can enter the Lower Spec Limit, Target, and Upper Spec Limit values for the process variable.

Figure 4.11 Capability Analysis Window

Lower Spec Limit 16.5
Target 21.5
Upper Spec Limit 23

OK Cancel Help

An example of a capability analysis report is shown in Figure 4.12 for Coating.jmp when the Lower Spec Limit is set as 16.5, the Target is set to 21.5, and the Upper Spec Limit is set to 23.

Figure 4.12 Capability Analysis Report for Coating.jmp

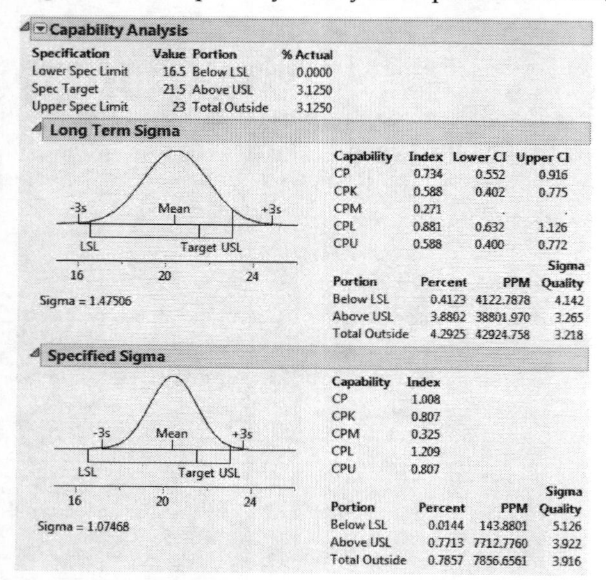

For additional information about Capability Analysis, see the Distributions chapter in the *Basic Analysis* book.

Save Sigma Saves the computed value of sigma as a column property in the process variable column in the JMP data table.

Save Limits > in Column Saves the computed values of sigma, center line, and the upper and lower limits as column properties in the process variable column in the JMP data table.

Save Limits > in New Table Saves all parameters for the particular chart type, including sigma and K Sigma, sample size, the center line, and the upper and lower control limits in a new JMP data table. Save this data table to use the limits later. On the Control Chart launch window, click **Get Limits** and then select the saved data table. See the section "Saving and Retrieving Limits" on page 93 for more information.

Save Summaries Creates a new data table that contains the sample label, sample sizes, the statistic being plotted, the center line, and the control limits. The specific statistics included in the table depend on the type of chart.

Alarm Script Enables you to write and run a script that indicates when the data fail special causes tests. Results can be written to the log or spoken. See "Tests" on page 48 in the "Control Chart Builder" chapter of this guide for more information. See the Scripting Guide for more information about writing custom Alarm Scripts.

See the JMP Reports chapter in the *Using JMP* book for more information about the following options:

Redo Contains options that enable you to repeat or relaunch the analysis. In platforms that support the feature, the Automatic Recalc option immediately reflects the changes that you make to the data table in the corresponding report window.

Save Script Contains options that enable you to save a script that reproduces the report to several destinations.

Save By-Group Script Contains options that enable you to save a script that reproduces the platform report for all levels of a By variable to several destinations. Available only when a By variable is specified in the launch window.

Individual Control Chart Options

The red triangle menu of chart options appears when you click the icon next to the chart name. Some options are also available under **Chart Options** when you right-click the chart.

Box Plots Superimposes box plots on the subgroup means plotted in a Mean chart. The box plot shows the subgroup maximum, minimum, 75th percentile, 25th percentile, and median. Markers for subgroup means show unless you deselect the **Show Points** option. The control limits displayed apply only to the subgroup mean. The **Box Plots** option is available only for $\bar{X}$-charts. It is most appropriate for larger subgroup sample sizes (more than 10 samples in a subgroup).

Needle Connects plotted points to the center line with a vertical line segment.

Connect Points Shows or hides the line that connects the data points.

Show Points Shows or hides the points representing summary statistics. Initially, the points show. You can use this option to suppress the markers denoting subgroup means when the **Box Plots** option is in effect.

Connect Color Displays the JMP color palette for you to choose the color of the line segments used to connect points.

Center Line Color Displays the JMP color palette for you to choose the color of the line segments used to draw the center line.

Limits Color Displays the JMP color palette for you to choose the color of the line segments used in the upper and lower limits lines.

Line Width Allows you to select the width of the control lines. Options are **Thin**, **Medium**, or **Thick**.

Point Marker Allows you to select the marker used on the chart.

Show Center Line Initially displays the center line in green. Deselecting **Show Center Line** removes the center line and its legend from the chart.

Show Control Limits Shows or hides the chart control limits and their legends.

Limits Precision Sets the decimal limit for labels.

Tests Shows a submenu that enables you to choose which tests to mark on the chart when the test is positive. Tests apply only for charts whose limits are 3σ limits. Tests 1 to 4 apply to Mean, Individual, and attribute charts. Tests 5 to 8 apply to Mean charts, Presummarize, and Individual Measurement charts only. If tests do not apply to a chart, the Tests option is dimmed. When sample sizes are unequal, the Test options are grayed out. If the samples change while the chart is open and they become equally sized, and the zone and/or test option is selected, the zones and/or tests are applied immediately and appear on the chart. These special tests are also referred to as the *Western Electric Rules*. For more information about special causes tests, see "Tests" on page 48 in the "Control Chart Builder" chapter.

Westgard Rules Westgard rules are control rules that help you decide whether a process is in or out of control. The different tests are abbreviated with the decision rule for the particular test. See the text and chart in "Westgard Rules" on page 51 in the "Control Chart Builder" chapter.

Test Beyond Limits Flags as a "*" any point that is beyond the limits. This test works on all charts with limits, regardless of the sample size being constant, and regardless of the size of *k* or the width of the limits. For example, if you had unequal sample sizes, and wanted to flag any points beyond the limits of an *r*-chart, you could use this command.

Show Zones Shows or hides the *zone lines.* The zones are labeled A, B, and C as shown here in the Mean plot for weight in the Coating.jmp sample data. Control Chart tests use the zone lines as boundaries. The seven zone lines are set one sigma apart, centered on the center line.

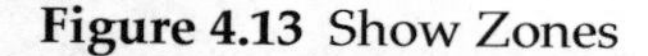

Figure 4.13 Show Zones

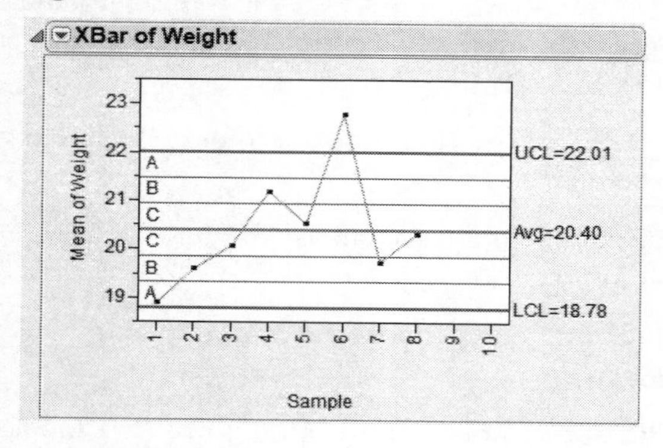

Shade Zones Shows or hides the default green, yellow, and red colors for the three zone areas and the area outside the zones. Green represents the area one sigma from the center line, yellow represents the area two and three sigmas from the center line, and red represents the area beyond three sigma. Shades can be shown with or without the zone lines.

Figure 4.14 Shade Zones

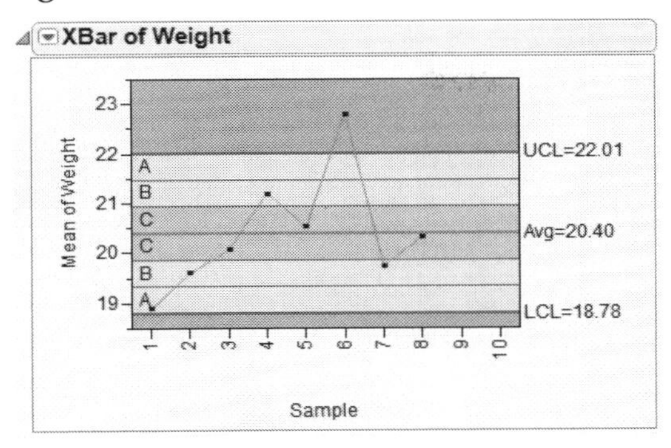

OC Curve Gives Operating Characteristic (OC) curves for specific control charts. OC curves are defined in JMP only for $\overline{X}$-, *p*-, *np*-, *c*-, and *u*-charts. The curve shows how the probability of accepting a lot changes with the quality of the sample. When you choose the **OC Curve** option from the control chart option list, JMP opens a new window containing the curve, using all the calculated values directly from the active control chart. Alternatively, you can run an OC curve directly from the **Control** category of the JMP Starter. Select the chart on which you want the curve based, then a window prompts you for **Target**, **Lower Control Limit**, **Upper Control Limit**, **k**, **Sigma**, and **Sample Size.** You can also perform both single and double acceptance sampling in the same manner. To engage this feature, choose **View** > **JMP Starter** > **Control** (under Click Category) > **OC Curves.** A pop-up window enables you to specify whether single or double acceptance sampling is desired. A second pop-up window is invoked, where you can specify acceptance failures, number inspected, and lot size (for single acceptance sampling). Clicking **OK** generates the desired OC curve.

Saving and Retrieving Limits

JMP can use previously established control limits for control charts:

- Upper and lower control limits, and a center line value.
- Parameters for computing limits such as a mean and standard deviation.

The control limits or limit parameter values must be either in a JMP data table, referred to as the *Limits Table*, or stored as a column property in the process column. When you specify the **Control Chart** command, you can retrieve the Limits Table with the **Get Limits** button on the Control Chart launch window.

The easiest way to create a Limits Table is to save results computed by the Control Chart platform. The **Save Limits** command in the red triangle menu for each control chart automatically saves limits from the sample values. The type of data saved in the table varies according to the type of control chart in the analysis window. You can also use values from any source and create your own Limits Table.

All Limits Tables must have:

- A column of special keywords that identify each row.
- A column for each of the variables whose values are the known standard parameters or limits. This column name must be the same as the corresponding process variable name in the data table to be analyzed by the Control Chart platform.

Table 3.10 on page 54 describes the limit keywords and their associated control chart.

You can save limits in a new data table or as properties of the response column. When you save control limits using the **in New Table** command, the limit keywords written to the table depend on the current chart types displayed.

Figure 4.15 shows examples of control limits saved to a data table using Coating.jmp. The rows with values _Mean, _LCL, and _UCL are for the Individual Measurement chart. The values with the R suffix (_AvgR, _LCLR, and _UCLR) are for the Moving Range chart. If you create these charts again using this Limits Table, the Control Chart platform identifies the appropriate limits from keywords in the _LimitsKey column.

Figure 4.15 Example of Saving Limits in a Data Table

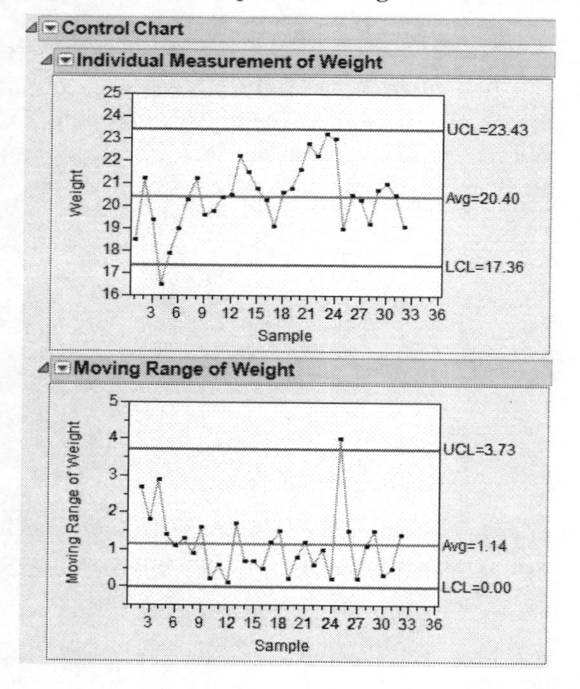

	_LimitsKey	Weight
1	_KSigma	3
2	_Alpha	0.0026997961
3	_Range Span	2
4	_Std Dev	1.0120139729
5	_Mean	20.396875
6	_LCL	17.360833081
7	_UCL	23.432916919
8	_AvgR	1.1419354839
9	_LCLR	0
10	_UCLR	3.7301687077

Note that values for _KSigma, _Alpha, and _Range Span can be specified in the Control Chart Launch window. JMP always looks at the values from the window first. Values specified in the window take precedence over those in an active Limits Table.

Rows with unknown keywords and rows marked with the excluded row state are ignored. Except for _Range Span, _KSigma, _Alpha, and _Sample Size, any needed values not specified are estimated from the data.

Excluded, Hidden, and Deleted Samples

The following table summarizes the effects of various conditions on samples and subgroups:

Table 4.1 Excluded, Hidden, and Deleted Samples

All rows of the sample are excluded before creating the chart.	Sample is not included in the calculation of the limits, but it appears on the graph.
Sample is excluded after creating the chart.	Sample is included in the calculation of the limits, and it appears in the graph. Nothing changes on the output by excluding a sample with the graph open.
Sample is hidden before creating the chart.	Sample is included in the calculation of the limits, but does not appear on the graph.
Sample is hidden after creating the chart.	Sample is included in the calculation of the limits, but does not appear on the graph. The sample marker disappears from the graph, the sample label still appears on the axis, but limits remain the same.
All rows of the sample are both excluded and hidden before creating the chart.	Sample is not included in the calculation of the limits, and it does not appear on the graph.
All rows of the sample are both excluded and hidden after creating the chart.	Sample is included in the calculation of the limits, but does not appear on the graph. The sample marker disappears from the graph, the sample label still appears on the axis, but limits remain the same.
Data set is subsetted with Sample deleted before creating chart.	Sample is not included in the calculation of the limits, the axis does not include a value for the sample, and the sample marker does not appear on the graph.

Table 4.1 Excluded, Hidden, and Deleted Samples *(Continued)*

Data set is subsetted with Sample deleted after creating chart.	Sample is not included in the calculation of the limits, and does not appear on the graph. The sample marker disappears from the graph, the sample label is removed from the axis, the graph shifts, and the limits change.

Some additional notes:

- Hide and Exclude operate only on the row state of the first observation in the sample. For example, if the second observation in the sample is hidden, while the first observation is not hidden, the sample will still appear on the chart.
- An exception to the exclude/hide rule: Both hidden and excluded rows are included in the count of points for Tests for Special Causes. An excluded row can be labeled with a special cause flag. A hidden point cannot be labeled. If the flag for a Tests for Special Causes is on a hidden point, it will not appear in the chart.
- Because of the specific rules in place (see Table 4.1 on page 95), the control charts do not support the Automatic Recalc script.

Additional Examples of the Control Chart Platform

This section contains additional examples using the Control Chart platform.

Run Chart Example

Run charts display a column of data as a connected series of points. The following example is a Run chart for the Weight variable from Coating.jmp in the Quality Control sample data folder (taken from the *ASTM Manual on Presentation of Data and Control Chart Analysis*).

1. Select **Help > Sample Data Library** and open Quality Control/Coating.jmp.
2. Select **Analyze > Quality and Process > Control Chart > Run Chart.**
3. Select Weight and click **Process.**
4. Select Sample and click **Sample Label.**
5. Click **OK.**

Figure 4.16 Run Chart

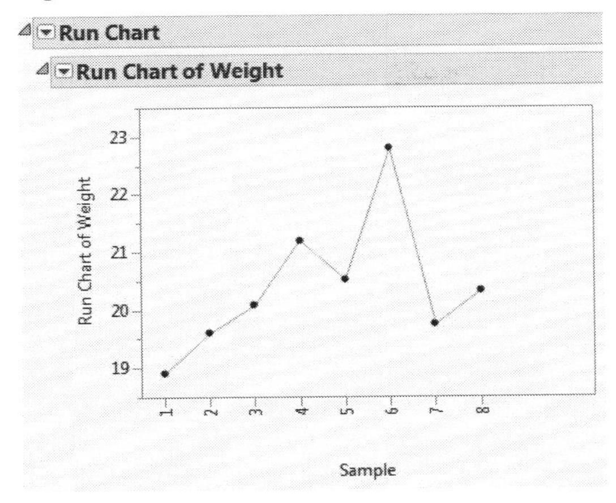

X Bar- and R-charts Example

The following example uses the Coating.jmp data table. The quality characteristic of interest is the Weight column. A subgroup sample of four is chosen. An $\overline{X}$-chart and an R-chart for the process are shown in Figure 4.17.

1. Select **Help > Sample Data Library** and open Quality Control/Coating.jmp.
2. Select **Analyze > Quality and Process > Control Chart > XBar.**

   Note the selected chart types of **XBar** and **R.**
3. Select Weight and click **Process.**
4. Select Sample and click **Sample Label.**
5. Click **OK.**

Sample six indicates that the process is not in statistical control. To check the sample values, click the sample six summary point on either control chart. The corresponding rows highlight in the data table.

Note: If an S chart is chosen with the $\overline{X}$-chart, then the limits for the $\overline{X}$-chart are based on the standard deviation. Otherwise, the limits for the $\overline{X}$-chart are based on the range.

Figure 4.17 Variables Charts for Coating Data

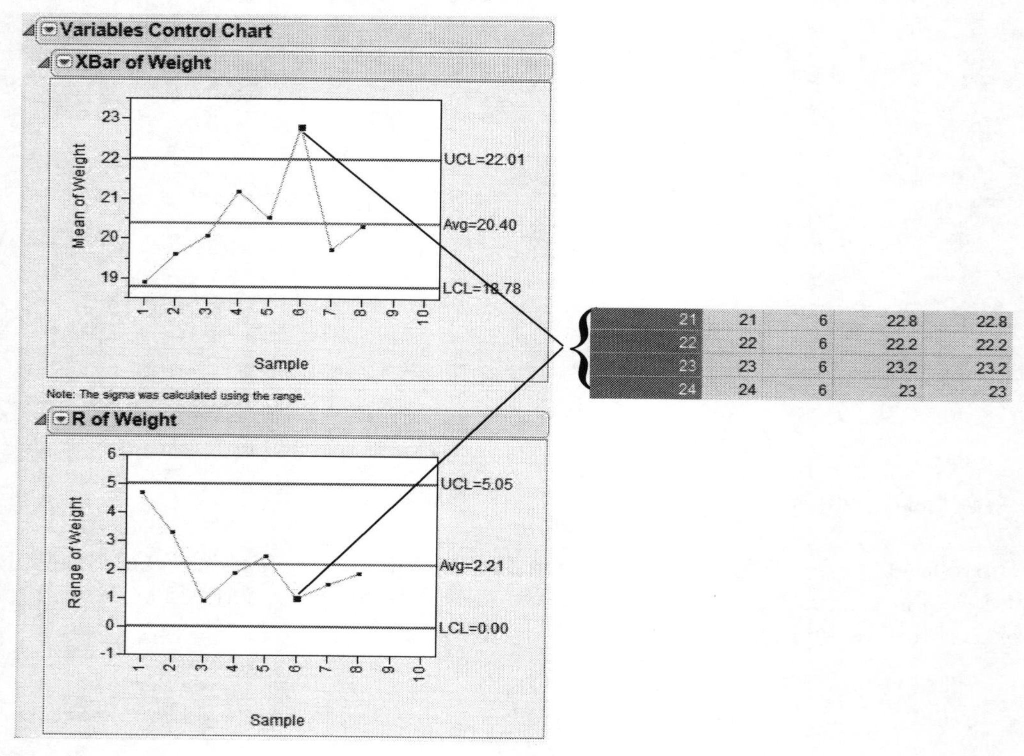

You can use **Fit Y by X** for an alternative visualization of the data. First, change the modeling type of Sample to Nominal. Specify the interval variable Weight as **Y, Response** and the nominal variable Sample as **X, Factor**. Select the **Quantiles** option from the Oneway Analysis drop-down menu. The box plots in Figure 4.18 show that the sixth sample has a small range of high values.

Figure 4.18 Quantiles Option in Fit Y By X Platform

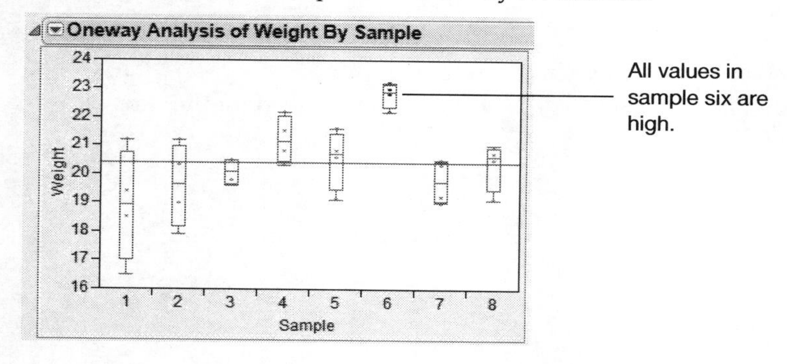

X-Bar- and S-charts with Varying Subgroup Sizes Example

The following example uses the Coating.jmp data table. This quality characteristic of interest is the Weight 2 column. An $\overline{X}$-chart and an S chart for the process are shown in Figure 4.19.

1. Select **Help > Sample Data Library** and open Quality Control/Coating.jmp.
2. Select **Analyze > Quality and Process > Control Chart > XBar.**
3. Select the chart types of **XBar** and **S.**
4. Select Weight 2 and click **Process.**
5. Select Sample and click **Sample Label.**

   The **Sample Size** option should automatically change to **Sample Grouped by Sample Label.**
6. Click **OK.**

Figure 4.19 $\overline{X}$ and S Charts for Varying Subgroup Sizes

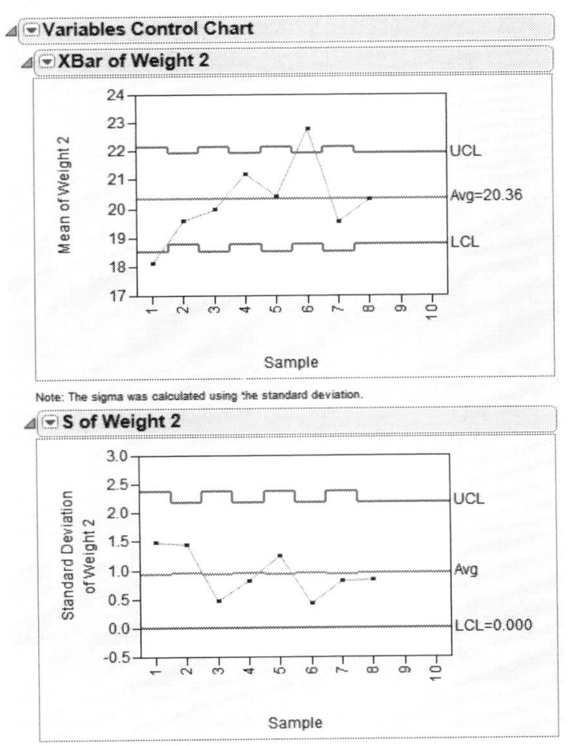

Weight 2 has several missing values in the data, so you might notice the chart has uneven limits. Although, each sample has the same number of observations, samples 1, 3, 5, and 7 each have a missing value.

Note: When sample sizes are unequal, the Test options are grayed out. If the samples change while the chart is open and they become equally sized, and the zone and/or test option is selected, the zones and/or tests will be applied immediately and appear on the chart.

Individual Measurement and Moving Range Charts Example

The Pickles.jmp data in the Quality Control sample data folder contains the acid content for vats of pickles. Because the pickles are sensitive to acidity and produced in large vats, high acidity ruins an entire pickle vat. The acidity in four vats is measured each day at 1, 2, and 3 PM. The data table records day, time, and acidity measurements. You can create Individual Measurement and Moving Range charts with date labels on the horizontal axis.

1. Select **Help > Sample Data Library** and open Quality Control/Pickles.jmp.
2. Select **Analyze > Quality and Process > Control Chart > IR.**
3. Select both **Individual Measurement** and **Moving Range** chart types.
4. Select Acid and click **Process.**
5. Select Date and click **Sample Label.**
6. Click **OK.**

The individual measurement and moving range charts shown in Figure 4.20 monitor the acidity in each vat produced.

Note: A Median Moving Range chart can also be evaluated. If you choose a Median Moving Range chart and an Individual Measurement chart, the limits on the Individual Measurement chart use the Median Moving Range as the sigma, rather than the Average Moving Range.

Figure 4.20 Individual Measurement and Moving Range Charts for Pickles Data

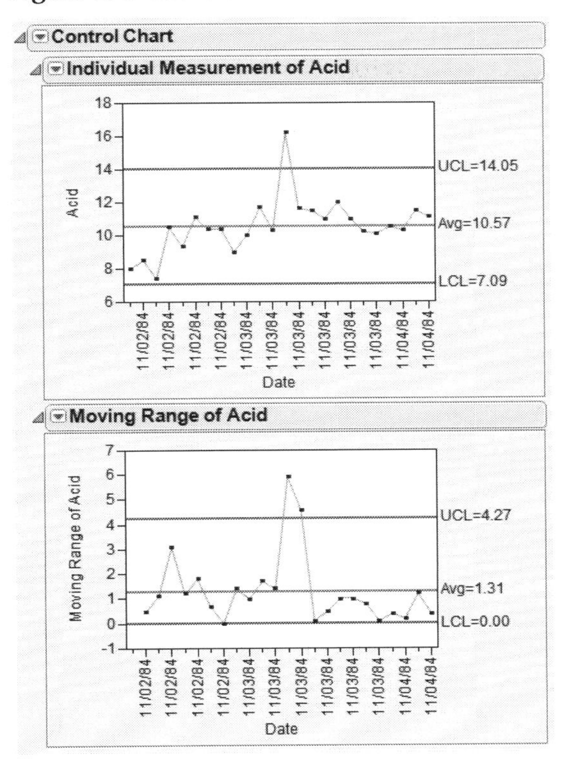

p-chart Example

Note: When you generate a *p*-chart and select **Capability**, JMP launches the Binomial Fit in Distribution and gives a Binomial-specific capability analysis.

The Washers.jmp data in the Quality Control sample data folder contains defect counts of 15 lots of 400 galvanized washers. The washers were inspected for finish defects such as rough galvanization and exposed steel. If a washer contained a finish defect, it was deemed nonconforming or defective. Thus, the defect count represents how many washers were defective for each lot of size 400. Using the Washers.jmp data table, specify a sample size variable, which would allow for varying sample sizes. This data contains all constant sample sizes.

1. Select **Help > Sample Data Library** and open Quality Control/Washers.jmp.
2. Select **Analyze > Quality and Process > Control Chart > P**.
3. Select # defective and click **Process.**
4. Select Lot and click **Sample Label.**

5. Select Lot Size and click **Sample Size.**
6. Click **OK.**

Figure 4.21 displays a *p*-chart for the proportion of defects.

Figure 4.21 p-chart

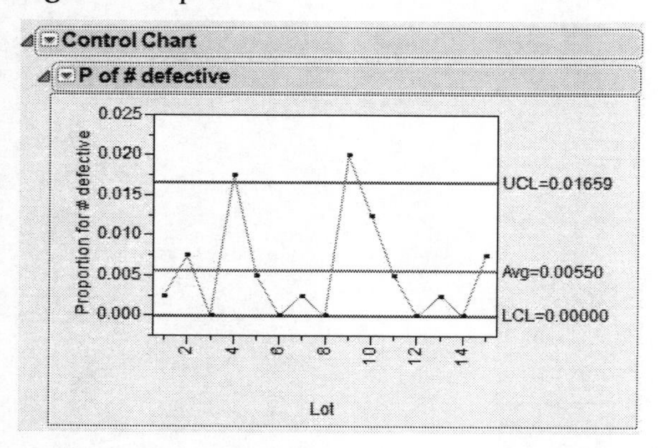

Note that although the points on the chart look the same as the *np*-chart in Figure 4.22, the *y* axis, Avg and limits are all different since they are now based on proportions.

np-chart Example

Note: When you generate a *np*-chart and select **Capability**, JMP launches the Binomial Fit in Distribution and gives a Binomial-specific capability analysis.

The following example uses the Washers.jmp data table.

- Select **Help > Sample Data Library** and open Quality Control/Washers.jmp.
- Select **Analyze > Quality and Process > Control Chart > NP**.
- Select # defective and click **Process.**
- Change the **Constant Size** to 400.
- Click **OK.**

Figure 4.22 displays an *np*-chart for the number of defects. Points 4 and 9 are above the upper control limit.

Figure 4.22 np-chart

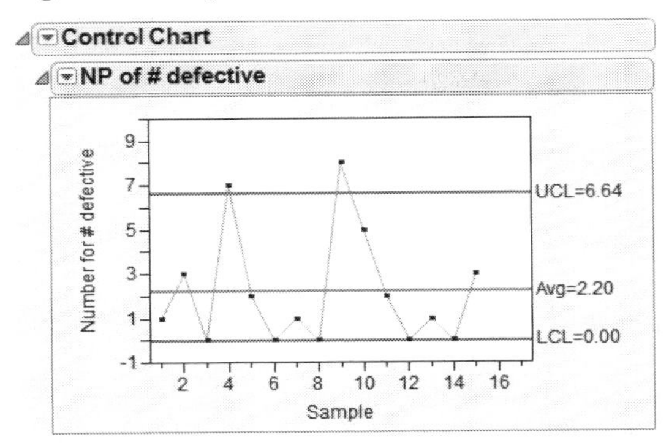

c-chart Example

c-charts are similar to *U*-charts in that they monitor the number of nonconformities in an entire subgroup, made up of one or more units. *c*-charts can also be used to monitor the average number of defects per inspection unit.

Note: When you generate a *c*-chart and select **Capability**, JMP launches the Poisson Fit in Distribution and gives a Poisson-specific capability analysis.

In this example, a clothing manufacturer ships shirts in boxes of ten. Prior to shipment, each shirt is inspected for flaws. Because the manufacturer is interested in the average number of flaws per shirt, the number of flaws found in each box is divided by ten and then recorded.

1. Select **Help > Sample Data Library** and open Quality Control/Shirts.jmp.
2. Select **Analyze > Quality and Process > Control Chart >** C.
3. Select # Defects and click **Process.**
4. Select Box and click **Sample Label.**
5. Select Box Size and click **Sample Size.**
6. Click **OK.**

Figure 4.23 c-chart

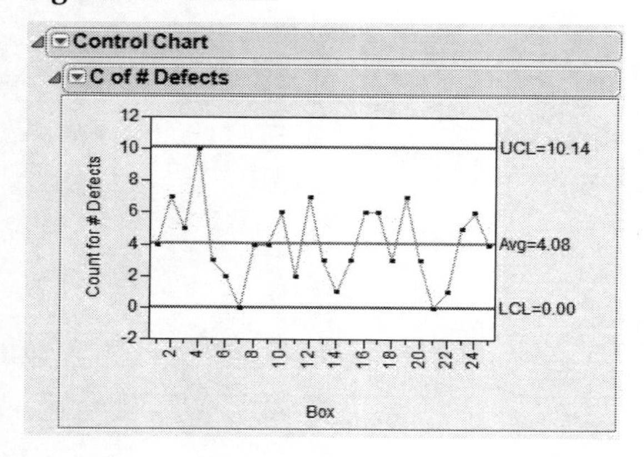

u-chart Example

The Braces.jmp data in the Quality Control sample data folder records the defect count in boxes of automobile support braces. A box of braces is one inspection unit. The number of boxes inspected (per day) is the subgroup sample size, which can vary. The *u*-chart in Figure 4.24 is monitoring the number of brace defects per subgroup sample size. The upper and lower bounds vary according to the number of units inspected.

Note: When you generate a *u*-chart, and select **Capability**, JMP launches the Poisson Fit in Distribution and gives a Poisson-specific capability analysis. To use the **Capability** feature, the unit sizes must be equal.

1. Select **Help > Sample Data Library** and open Quality Control/Braces.jmp.
2. Select **Analyze > Quality and Process > Control Chart > U**.
3. Select # defects and click **Process**.
4. Select Date and click **Sample Label**.
5. Select Unit size and click **Unit Size**.
6. Click **OK**.

Figure 4.24 u-chart

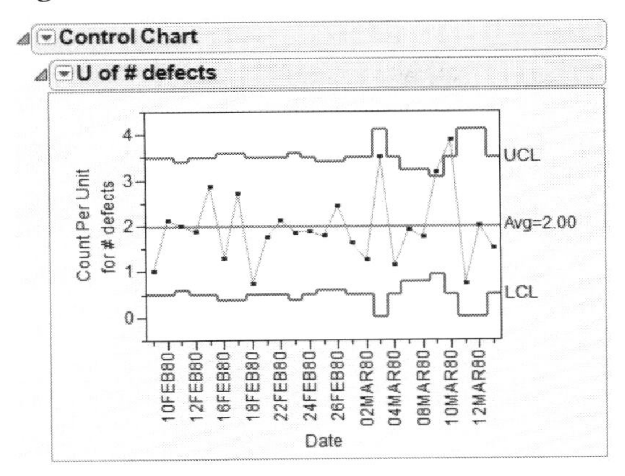

UWMA Chart Example

In sample data table, Clips1.jmp, the measure of interest is the gap between the ends of manufactured metal clips. To monitor the process for a change in average gap, subgroup samples of five clips are selected daily. A UWMA chart with a moving average span of three is examined.

1. Select **Help > Sample Data Library** and open Quality Control/Clips1.jmp.
1. Select **Analyze > Quality and Process > Control Chart > UWMA**.
2. Select Gap and click **Process**.
3. Select Sample and click **Sample Label**.
4. Change the **Moving Average Span** to 3.
5. Click **OK**.

The result is the chart in Figure 4.25. The point for the first day is the mean of the five subgroup sample values for that day. The plotted point for the second day is the average of subgroup sample means for the first and second days. The points for the remaining days are the average of subsample means for each day and the two previous days.

The average clip gap appears to be decreasing, but no sample point falls outside the 3σ limits.

Figure 4.25 UWMA Charts for the Clips1 data

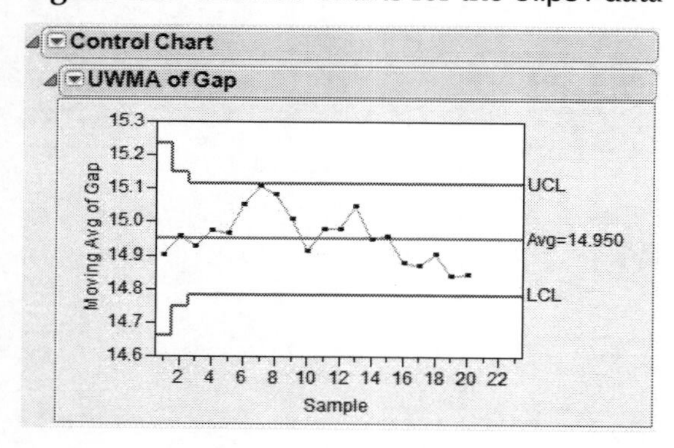

EWMA Chart Example

The following example uses the Clips1.jmp data table.

1. Select **Help > Sample Data Library** and open Quality Control/Clips1.jmp.
2. Select **Analyze > Quality and Process > Control Chart > EWMA.**
3. Select Gap and click **Process.**
4. Select Sample and click **Sample Label.**
5. Change the **Weight** to 0.5.
6. Leave the **Sample Size Constant** as 5.
7. Click **OK.**

Figure 4.26 displays the EWMA chart for the same data seen in Figure 4.25. This EWMA chart was generated for weight = 0.5.

Figure 4.26 EWMA Chart

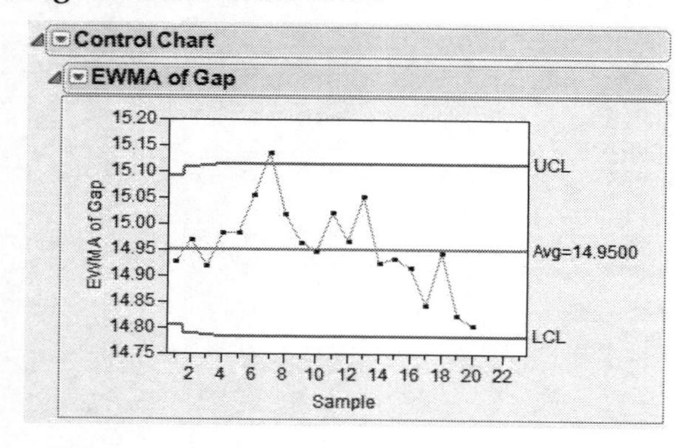

Presummarize Chart Example

The following example uses the Coating.jmp data table.

1. Select **Help > Sample Data Library** and open Quality Control/Coating.jmp.
2. Select **Analyze > Quality and Process > Control Chart > Presummarize.**
3. Select Weight and click **Process.**
4. Select Sample and click **Sample Label**.
5. Select both **Individual on Group Means** and **Moving Range on Group Means.** The **Sample Grouped by Sample Label** button is automatically selected when you choose a Sample Label variable.

   When using **Presummarize** charts, you can select either **On Group Means** options or **On Group Std Devs** options or both. Each option creates two charts (an Individual Measurement, also known as an X chart, and a Moving Range chart) if both IR chart types are selected.

   The **On Group Means** options compute each sample mean and then plot the means and create an Individual Measurement and a Moving Range chart on the means.

   The **On Group Std Devs** options compute each sample standard deviation and plot the standard deviations as individual points. Individual Measurement and Moving Range charts for the standard deviations then appear.
6. Click **OK**.

Figure 4.27 Example of Charting Presummarized Data

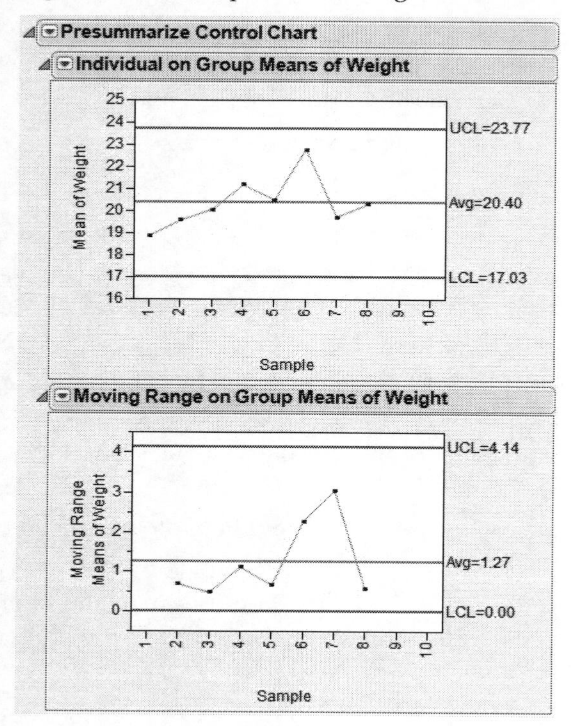

Although the points for $\overline{X}$- and *S*-charts are the same as the Individual on Group Means and Individual on Group Std Devs charts, the limits are different because they are computed as Individual charts.

Another way to generate the presummarized charts, with the Coating.jmp data table:

1. Choose **Tables > Summary**.
2. Assign Sample as the **Group** variable, then Mean(Weight) and Std Dev(Weight) as **Statistics**.
3. Click **OK**.
4. Select **Analyze > Quality and Process > Control Chart > IR**.
5. Select Mean(Weight) and Std Dev(Weight) and click **Process**.
6. Click **OK**.

The resulting charts match the presummarized charts.

Phase Example

Open Diameter.jmp, found in the Quality Control sample data folder. This data set contains the diameters taken for each day, both with the first prototype (phase 1) and the second prototype (phase 2).

- Select **Help > Sample Data Library** and open Quality Control/Diameter.jmp.
- Select **Analyze > Quality and Process > Control Chart > XBar**.
- Select DIAMETER and click **Process**.
- Select DAY and click **Sample Label**.
- Select Phase and click **Phase**.
- Select **S** and **XBar**.
- Click **OK**.

The resulting chart has different limits for each phase.

Figure 4.28 Phase Control Chart

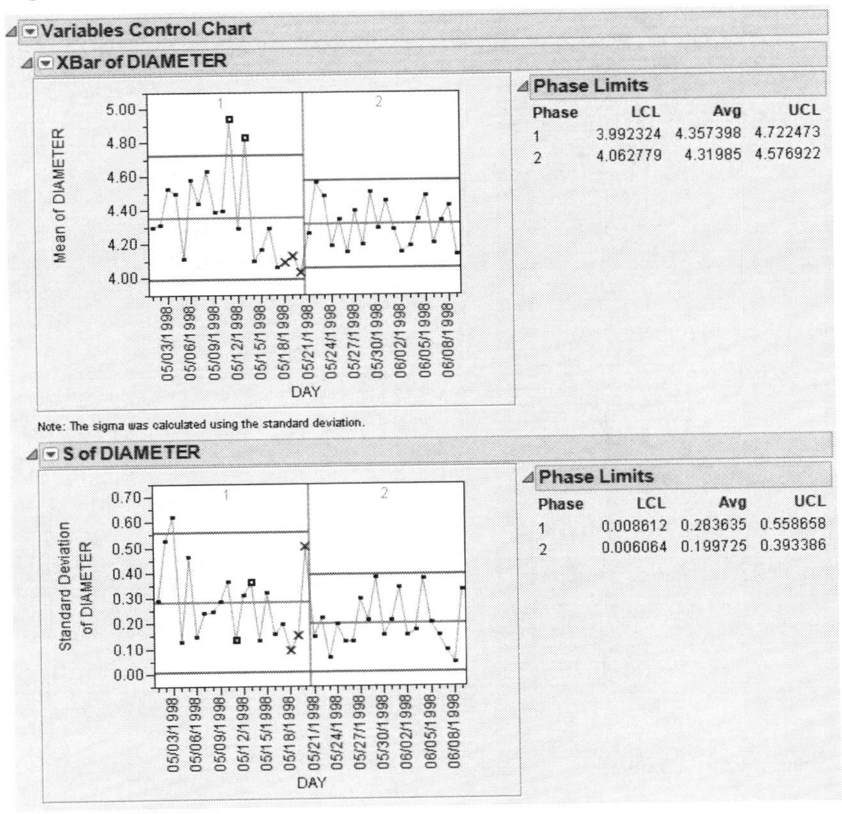

Statistical Details for the Control Chart Platform

This section contains statistical details for median moving range charts, UWMA charts, and EWMA charts. For details on any other types of charts (such as $\overline{X}$- and R-charts, p- and np-charts, and more) see the "Statistical Details for the Control Chart Builder Platform" on page 67 in the "Control Chart Builder" chapter.

Control Limits for Median Moving Range Charts

LCL for Median Moving Range Chart = max(0, MMR - (k*Std Dev*$d_3(n)$))

UCL for Median Moving Range Chart = MMR + (k*Std Dev*$d_3(n)$)

The standard deviation for Median Moving Range charts is estimated by:

Std Dev = MMR/$d_4(n)$

where:

MMR = Center Line (Avg) for Median Moving Range chart

$d_4(n)$ = expected value of the range of a normally distributed sample of size n.

Control Limits for UWMA Charts

Control limits for UWMA charts are computed as follows. For each subgroup i,

$$\text{LCL}_i = \overline{X}_w - k\frac{\hat{\sigma}}{min(i, w)}\sqrt{\frac{1}{n_i} + \frac{1}{n_{i-1}} + \ldots + \frac{1}{n_{1+max(i-w, 0)}}}$$

$$\text{UCL}_i = \overline{X}_w + k\frac{\hat{\sigma}}{min(i, w)}\sqrt{\frac{1}{n_i} + \frac{1}{n_{i-1}} + \ldots + \frac{1}{n_{1+max(i-w, 0)}}}$$

where:

w is the span parameter (number of terms in moving average)

n_i is the sample size of the ith subgroup

k is the number of standard deviations

$\overline{X}_w$ is the weighted average of subgroup means

$\hat{\sigma}$ is the process standard deviation

Control Limits for EWMA Charts

Control limits for EWMA charts are computed as:

$$\text{LCL} = \overline{X}_w - k\hat{\sigma}r\sqrt{\sum_{j=0}^{i-1}\frac{(1-r)^{2j}}{n_{i-j}}}$$

$$\text{UCL} = \overline{X}_w + k\hat{\sigma}r\sqrt{\sum_{j=0}^{i-1}\frac{(1-r)^{2j}}{n_{i-j}}}$$

where:

r is the EWMA weight parameter ($0 < r \leq 1$)

x_{ij} is the jth measurement in the ith subgroup, with $j = 1, 2, 3, \ldots, n_i$

n_i is the sample size of the ith subgroup

k is the number of standard deviations

$\overline{X}_w$ is the weighted average of subgroup means

$\hat{\sigma}$ is the process standard deviation

Chapter 5

Cumulative Sum Control Charts

Detect Small Shifts in the Process Mean

CUSUM charts show cumulative sums of subgroup or individual measurements from a target value. CUSUM charts can help you decide whether a process is in a state of statistical control by detecting small, sustained shifts in the process mean. In comparison, Shewhart charts can detect sudden and large changes in measurement, such as a two or three sigma shift, but they are ineffective at spotting smaller changes, such as a one sigma shift.

Figure 5.1 Example of a CUSUM Chart

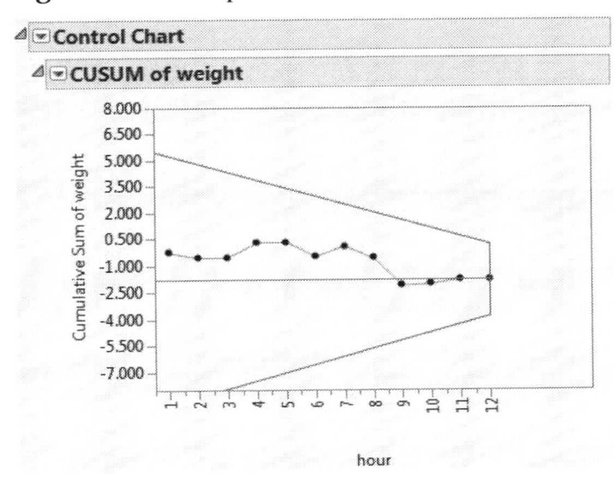

CUSUM Control Chart Overview

Cumulative Sum (CUSUM) control charts show cumulative sums of subgroup or individual measurements from a target value. CUSUM charts can help you decide whether a process is in a state of statistical control by detecting small, sustained shifts in the process mean. In comparison, Shewhart charts can detect sudden and large changes in measurement, such as a two or three sigma shift, but they are ineffective at spotting smaller changes, such as a one sigma shift.

Example of a CUSUM Chart

A machine fills 8-ounce cans of two-cycle engine oil additive. The filling process is believed to be in statistical control. The process is set so that the average weight of a filled can (μ_0) is 8.10 ounces. Previous analysis shows that the standard deviation of fill weights (σ_0) is 0.05 ounces.

Subgroup samples of four cans are selected and weighed every hour for twelve hours. Each observation in the Oil1 Cusum.jmp data table contains one value of weight and its associated value of hour. The observations are sorted so that the values of hour are in increasing order.

1. Select **Help > Sample Data Library** and open Quality Control/Oil1 Cusum.jmp.
2. Select **Analyze > Quality And Process > Control Chart > CUSUM.**
3. Select weight and click **Process.**
4. Select hour and click **Sample Label.**
5. Select the **Two Sided** check box if it is not already checked.
6. In the Parameters area, click the **H** button and type 2.
7. Click **Specify Stats.**
8. Type 8.1 next to **Target.**

   8.1 is the average weight in ounces of a filled can. This is the target mean.
9. Type 1 next to **Delta.**

   1 is the absolute value of the smallest shift to be detected as a multiple of the process standard deviation or of the standard error.
10. Type 0.05 next to **Sigma.**

    0.05 is the known standard deviation of fill weights (σ_0) in ounces.

Figure 5.2 Completed Launch Window

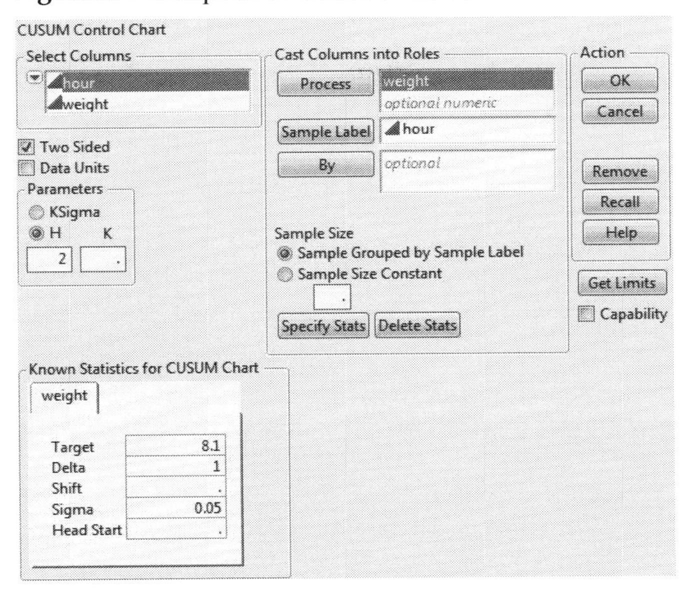

11. Click **OK**.

Figure 5.3 Two-Sided CUSUM Chart for Oil1 Cusum.jmp Data

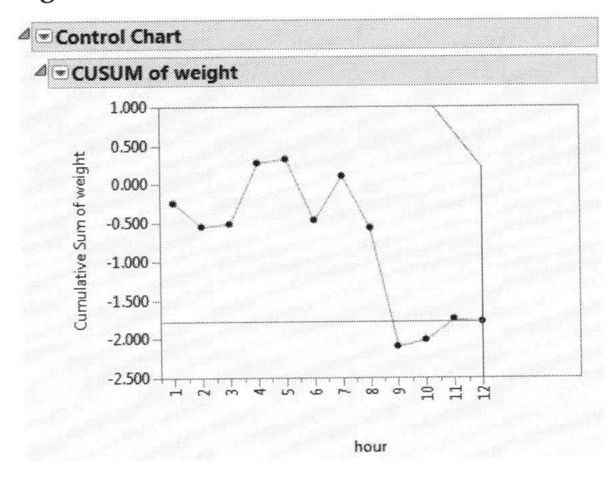

You can interpret the chart by comparing the points with the V-mask. The right edge of the V-mask is centered at the most recent point (the 12th hour). Because none of the points cross the arms of the V-mask, there is no evidence that a shift in the process has occurred. See "Interpret a Two-Sided CUSUM Chart" on page 118.

Launch the CUSUM Control Chart Platform

Launch the CUSUM Control Chart platform by selecting **Analyze > Quality And Process > Control Chart > CUSUM**.

Figure 5.4 The CUSUM Control Chart Launch Window

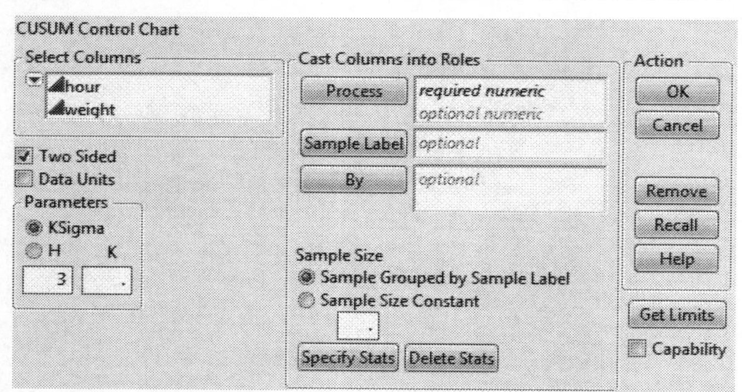

Process Identifies the variables that you want to chart. For variables charts, specify measurements as the process. For attribute charts, specify the defect count or defective proportion as the process. The data are interpreted as counts, unless it contains non-integer values between 0 and 1.

Note: The rows of the data table must be sorted in the order in which you want them to appear in the control chart. Even if there is a Sample Label variable specified, you still must sort the data accordingly.

Sample Label Specify a variable whose values label the horizontal axis and can also identify unequal subgroup sizes. If no sample label variable is specified, the samples are identified by their subgroup sample number. See "Sample Label" on page 82 in the "Shewhart Control Charts" chapter.

By Identifies a column that creates a report consisting of separate analyses for each level of the variable.

Two Sided Requests a two-sided CUSUM chart when selected. If it is not selected, a one-sided chart is used and no V-mask appears. If an H value is specified, a decision interval is displayed.

Data Units Specifies that the cumulative sums be computed without standardizing the subgroup means or individual values. The vertical axis of the CUSUM chart is then scaled in the same units as the data.

Note: Data Units requires that the subgroup sample size be designated as constant.

K Sigma Specifies control limits in terms of a multiple of the sample standard error. Enter the K Sigma value in the box that appears below H. Control limits are specified at k sample standard errors above and below the expected value, which shows as the shift. See "KSigma" on page 85 in the "Shewhart Control Charts" chapter.

H H is the vertical distance *h* between the origin for the V-mask and the upper or lower arm of the V-mask for a two-sided chart (for an illustration, see Figure 5.6). For a one-sided chart, H is the decision interval. Choose H as a multiple of the standard error.

K Specifies reference value k, where k is greater than zero.

Sample Grouped by Sample Label Indicates the column that identifies each sample. See "Sample Label" on page 82 in the "Shewhart Control Charts" chapter.

Sample Size Constant Indicates that the sample subgroups are the same size. See "Sample Label" on page 82 in the "Shewhart Control Charts" chapter.

Specify Stats Enter the following process variable specifications in the Known Statistics for CUSUM Chart area:

- **Target** is the target mean (goal) for the process or population. The target mean must be scaled in the same units as the data.
- **Delta** specifies the absolute value of the smallest shift to be detected as a multiple of the process standard deviation or of the standard error. This depends on whether the shift is viewed as a shift in the population mean or as a shift in the sampling distribution of the subgroup mean, respectively. Delta is an alternative to the Shift option (described next). The relationship between Shift and Delta can be computed as follows:

  $$\delta = \frac{\Delta}{(\sigma/(\sqrt{n}))}$$

  where δ represents Delta, Δ represents the shift, σ represents the process standard deviation, and n is the (common) subgroup sample size.
- **Shift** is the minimum value that you want to detect on either side of the target mean. You enter the shift value in the same units as the data, and you interpret it as a shift in the mean of the sampling distribution of the subgroup mean. You can choose either Shift or Delta.
- **Sigma** specifies a known standard deviation, σ_0, for the process standard deviation, σ. By default, the Control Chart platform estimates sigma from the data.
- **Head Start** specifies an initial value for the cumulative sum, S_0, for a one-sided CUSUM chart (S_0 is usually zero). Enter the **Head Start** value as a multiple of standard error.

Delete Stats Deletes all statistics in the Known Statistics for CUSUM Chart area.

Get Limits Uses previously established limits that exist in a JMP data table. See "Saving and Retrieving Limits" on page 93 in the "Shewhart Control Charts" chapter.

Capability Measures the conformance of a process to given specification limits. Once you click **OK** in the launch window, if you have not already defined these as a column property, you are prompted to enter specification limits and a target. See the Distributions chapter in the *Basic Analysis* book.

For more information about the launch window, see the Get Started chapter in the *Using JMP* book.

The CUSUM Control Chart

CUSUM charts can be one-sided or two-sided. One-sided charts detect a shift in one direction from a specified target mean. Two-sided charts detect a shift in either direction.

Figure 5.5 Example of a Two-Sided CUSUM Chart

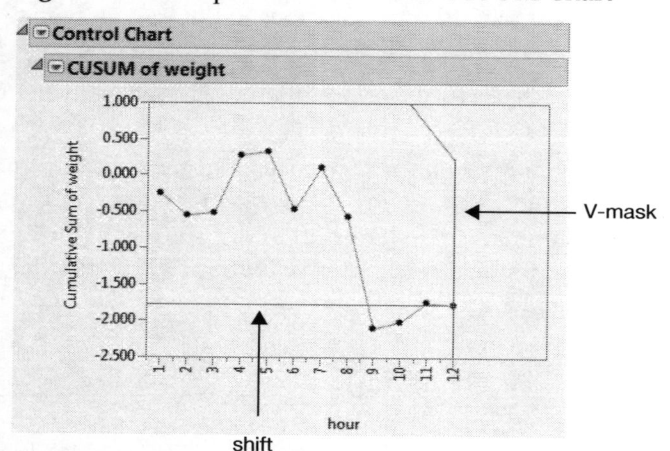

Note the following:

- If you use the grabber tool and click on a point, the shift and V-mask adjust to reflect the process condition at that point.
- If you add new data to the data table for an existing CUSUM chart, the corresponding chart updates automatically.

For information about additional options, see "CUSUM Control Chart Platform Options" on page 120.

Interpret a Two-Sided CUSUM Chart

To interpret a two-sided CUSUM chart, compare the points with limits that compose a V-mask. A V-mask is a shape in the form of a V on its side that is superimposed on the graph of the cumulative sums. The V-mask is formed by plotting V-shaped limits. The origin of a V-mask is the most recently plotted point, and the arms extended backward on the x-axis, as in Figure 5.6. As data are collected, the cumulative sum sequence is updated and the origin is relocated at the newest point.

Figure 5.6 V-Mask for a Two-Sided CUSUM Chart

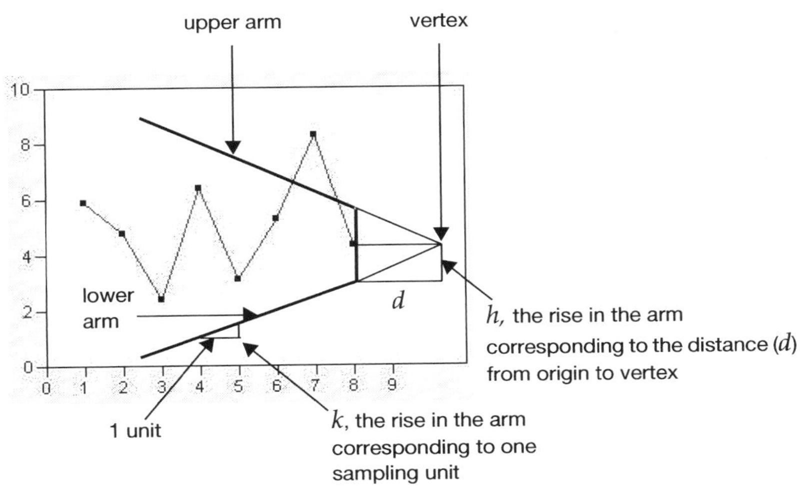

Shifts in the process mean are visually easy to detect on a CUSUM chart because they produce a change in the slope of the plotted points. The point where the slope changes is the point where the shift occurs. A condition is *out-of-control* if one or more of the points previously plotted crosses the upper or lower arm of the V-mask. Points crossing the lower arm signal an increasing process mean, and points crossing the upper arm signal a downward shift.

There are important differences between CUSUM charts and Shewhart charts:

- A Shewhart control chart plots points based on information from a single subgroup sample. In CUSUM charts, each point is based on information from all samples taken up to and including the current subgroup.
- On a Shewhart control chart, horizontal control limits define whether a point signals an out-of-control condition. On a CUSUM chart, the limits can be either in the form of a V-mask or a horizontal decision interval.
- The control limits on a Shewhart control chart are commonly specified as 3σ limits. On a CUSUM chart, the limits are determined from average run length, from error probabilities, or from an economic design.

A CUSUM chart is more efficient for detecting small shifts in the process mean. Lucas (1976) says that a V-mask detects a 1σ shift about four times as fast as a Shewhart control chart.

Interpret a One-Sided CUSUM Chart

Use a one-sided CUSUM chart to identify data approaching or exceeding the side of interest.

Figure 5.7 Example of a One-Sided CUSUM Chart

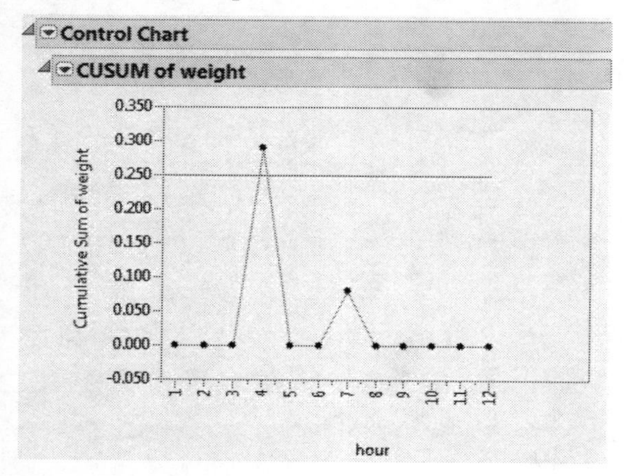

The *decision interval* or horizontal line is set at the H value that you entered in the launch window. In this example, it is 0.25. Any values exceeding the decision interval of 0.25 indicate a shift or out-of-control condition. In this example, observation 4 appears to be where a shift occurred. Also note that no V-mask appears for one-sided CUSUM charts.

CUSUM Control Chart Platform Options

For a description of the options in the red triangle menu next to Control Chart, see "Control Chart Window Options" on page 88 in the "Shewhart Control Charts" chapter. The red triangle menu next to CUSUM contains the following options:

Show Points Shows or hides the sample data points.

Connect Points Connects the sample points with a line.

Mask Color (Applicable only when Show V Mask is selected) Select a line color for the V-mask.

Connect Color (Applicable only when Connect Points is selected) Select a color for the connect line.

Center Line Color (Applicable only when Show Shift is selected) Select a color for the center line, or shift.

Show Shift Shows or hides the shift that you entered in the launch window.

Show V Mask Shows or hides the V-mask based on the statistics that you specified in the CUSUM Control Charts launch window.

Show Parameters Shows or hides a report that summarizes the CUSUM charting parameters.

Show ARL Shows or hides the average run length (ARL) information. The average run length is the expected number of samples taken before an out-of-control condition is signaled, as follows:

- ARL (Delta), sometimes denoted ARL1, is the average run length for detecting a shift in the size of the specified Delta.
- ARL(0), sometimes denoted ARL0, is the in-control average run length for the specified parameters (Montgomery 2013).

Example of a One-Sided CUSUM Chart

Consider the data used in "Example of a CUSUM Chart" on page 114, where the machine fills 8-ounce cans of engine oil. In order to cut costs, the manufacturer is now concerned about significant over-filling (and not so concerned about under-filling). Use a one-sided CUSUM chart to identify any instances of over-filling. Anything that is 0.25 ounces beyond the mean of 8.1 is considered a problem.

1. Select **Help > Sample Data Library** and open Quality Control/Oil1 Cusum.jmp.
2. Select **Analyze > Quality And Process > Control Chart > CUSUM**.
3. Deselect **Two Sided**.
4. Select weight and click **Process**.
5. Select hour and click **Sample Label**.
6. Click **H** and type 0.25.
7. Click **Specify Stats**.
8. Type 8.1 next to **Target**.

   8.1 is the average weight in ounces of a filled can. This is the target mean.
9. Type 1 next to **Delta**.

   1 is the absolute value of the smallest shift to be detected as a multiple of the process standard deviation or of the standard error.
10. Type 0.05 next to **Sigma**.

    0.05 is the known standard deviation of fill weights (σ_0) in ounces.
11. Click **OK**.

Figure 5.8 One-Sided CUSUM Chart for Oil1 Cusum.jmp Data

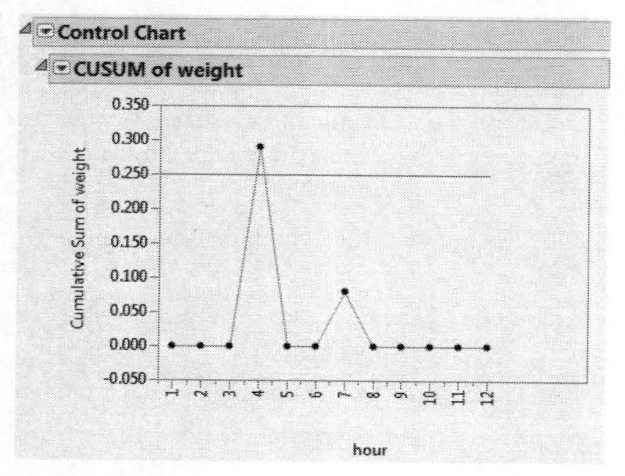

The decision interval is set at the H value that you entered (0.25). You can see that at the fourth hour, some significant over-filling occurred.

Statistical Details for CUSUM Control Charts

The following notation is used in these formulas:

- μ denotes the mean of the population, also referred to as the process mean or the process level.
- μ_0 denotes the target mean (or goal) for the population. Sometimes, the symbol $\overline{X}_0$ is used for μ_0. See the American Society for Quality Statistics Division (2004). You can provide μ_0 as the Target in the Known Statistics for CUSUM Chart area on the launch window.
- σ denotes the population standard deviation. $\hat{\sigma}$ denotes an estimate of σ.
- σ_0 denotes a known standard deviation. You can provide σ_o as the Sigma in the Known Statistics for CUSUM Chart area on the launch window.
- n denotes the nominal sample size for the CUSUM chart.
- δ denotes the shift in μ to be detected, expressed as a multiple of the standard deviation. You can provide δ as the Delta in the Known Statistics for CUSUM Chart area on the launch window.
- Δ denotes the shift in μ to be detected, expressed in data units. If the sample size n is constant across subgroups, then the following computation applies:

  $$\Delta = \delta\sigma_{\overline{X}} = (\delta\sigma)/\sqrt{n}$$

  You can provide Δ as the Shift in the Known Statistics for CUSUM Chart area on the launch window.

Note: Some authors use the symbol D instead of Δ.

One-Sided CUSUM Charts

Positive Shifts

If the shift δ to be detected is positive, the CUSUM for the tth subgroup is computed as follows:

$$S_t = \max(0, S_{t-1} + (z_t - k))$$

$t = 1, 2,..., n$, where $S_0 = 0$, z_t is defined as for two-sided charts, and the parameter k, termed the *reference value*, is positive. The CUSUM S_t is referred to as an *upper cumulative sum*. S_t can be computed as follows:

$$\max\left(0, S_{t-1} + \frac{\overline{X}_i - (\mu_0 + k\sigma_{\overline{X}_i})}{\sigma_{\overline{X}_i}}\right)$$

The sequence S_t cumulates deviations in the subgroup means greater than k standard errors from μ_0. If S_t exceeds a positive value h (referred to as the *decision interval*), a shift or out-of-control condition is signaled.

Negative Shifts

If the shift to be detected is negative, the CUSUM for the tth subgroup is computed as follows:

$$S_t = \max(0, S_{t-1} - (z_t + k))$$

$t = 1, 2,..., n$, where $S_0 = 0$, z_t is defined as for two-sided charts, and the parameter k, termed the *reference value*, is positive. The CUSUM S_t is referred to as a *lower cumulative sum*. S_t can be computed as follows:

$$\max\left(0, S_{t-1} - \frac{\overline{X}_i - (\mu_0 - k\sigma_{\overline{X}_i})}{\sigma_{\overline{X}_i}}\right)$$

The sequence S_t cumulates the absolute value of deviations in the subgroup means less than k standard errors from μ_0. If S_t exceeds a positive value h (referred to as the *decision interval*), a shift or out-of-control condition is signaled.

Note that S_t is always positive and h is always positive, regardless of whether δ is positive or negative. For charts designed to detect a negative shift, some authors define a reflected version of S_t for which a shift is signaled when S_t is less than a negative limit.

Lucas and Crosier (1982) describe the properties of a fast initial response (FIR) feature for CUSUM charts in which the initial CUSUM S_0 is set to a "head start" value. Average run length calculations given by them show that the FIR feature has little effect when the process is in control and that it leads to a faster response to an initial out-of-control condition than a standard CUSUM chart. You can provide a Head Start value in the Known Statistics for CUSUM Chart area on the launch window.

Constant Sample Sizes

When the subgroup sample sizes are constant (= n), it might be preferable to compute CUSUMs that are scaled in the same units as the data. CUSUMs are then computed as follows:

$$S_t = \max(0, S_{t-1} + (\overline{X}_i - (\mu_0 + k\sigma/\sqrt{n})))$$

where $\delta > 0$

$$S_t = \max(0, S_{t-1} - (\overline{X}_i - (\mu_0 - k\sigma/\sqrt{n})))$$

where $\delta < 0$. In either case, a shift is signaled if S_t exceeds $h' = h\sigma/\sqrt{n}$. Some authors use the symbol H for h'.

Two-Sided CUSUM Charts

If the CUSUM chart is two-sided, the cumulative sum S_t plotted for the tth subgroup is as follows:

$$S_t = S_{t-1} + z_t$$

$t = 1, 2,..., n$. Here S_0=0, and the term z_t is calculated as follows:

$$z_t = (\overline{X}_t - \mu_0)/(\sigma/\sqrt{n_t})$$

where $\overline{X}_t$ is the tth subgroup average, and n_t is the tth subgroup sample size. If the subgroup samples consist of individual measurements x_t, the term z_t simplifies to the following computation:

$$z_t = (x_t - \mu_0)/\sigma$$

The first equation can be rewritten as follows:

$$S_t = \sum_{i=1}^{t} z_i = \sum_{i=1}^{t} (\overline{X}_i - \mu_0)/\sigma_{\overline{X}_i}$$

where the sequence S_t cumulates standardized deviations of the subgroup averages from the target mean μ_0.

In many applications, the subgroup sample sizes n_i are constant ($n_i = n$), and the equation for S_t can be simplified, as follows:

$$S_t = (1/\sigma_{\overline{X}}) \sum_{i=1}^{t} (\overline{X}_i - \mu_0) = (\sqrt{n}/\sigma) \sum_{i=1}^{t} (\overline{X}_i - \mu_0)$$

In some applications, it might be preferable to compute S_t as follows:

$$S_t = \sum_{i=1}^{t} (\overline{X}_i - \mu_0)$$

which is scaled in the same units as the data. In this case, the procedure rescales the V-mask parameters h and k to $h' = h\sigma/\sqrt{n}$ and $k' = k\sigma/\sqrt{n}$, respectively. Some authors use the symbols F for k' and H for h'.

If the process is in control and the mean μ is at or near the target μ_0, the points will not exhibit a trend since positive and negative displacements from μ_0 tend to cancel each other. If μ shifts in the positive direction, the points exhibit an upward trend, and if μ shifts in the negative direction, the points exhibit a downward trend.

Chapter 6

Multivariate Control Charts

Monitor Multiple Process Characteristics Simultaneously

Univariate control charts monitor a single process characteristic, whereas multivariate control charts monitor multiple process characteristics. Independent variables can be charted individually, but if the variables are correlated, a multivariate chart is needed to determine whether the process is in control. Multivariate control charts can detect shifts in the mean or the relationship between several related variables.

Figure 6.1 Example of a Multivariate Control Chart

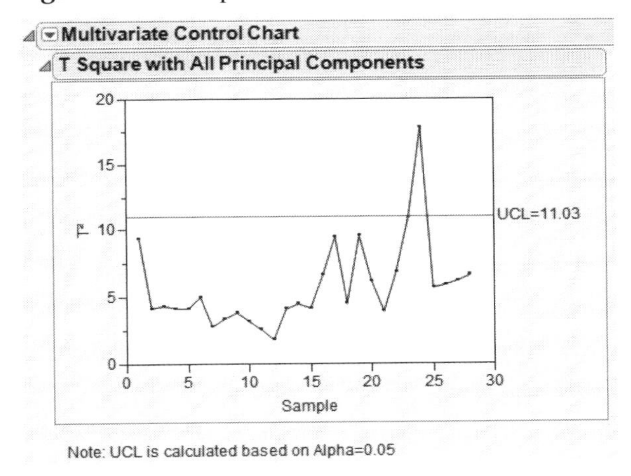

Multivariate Control Chart Overview

Univariate control charts monitor a single process characteristic, whereas multivariate control charts monitor multiple process characteristics. Independent variables can be charted individually, but if the variables are correlated, a multivariate chart is needed to determine whether the process is in control. Multivariate control charts can detect shifts in the mean or the relationship between several related variables.

To construct a multivariate control chart for individual or sub-grouped measurements, first identify a period of time during which the process is stable and capable. Then, using JMP, proceed as follows:

1. Develop a control chart to verify that the process is stable over this period.
2. Save the target statistics for this data.
3. Monitor the on-going process using a control chart based on these saved target statistics.

Example of a Multivariate Control Chart

The following example illustrates constructing a control chart for data that are not sub-grouped. The data are measurements on a steam turbine engine. For an example that uses sub-grouped data, "Example of Monitoring a Process Using Sub-Grouped Data" on page 136.

Step 1: Determine Whether the Process is Stable

1. Select **Help > Sample Data Library** and open Quality Control/Steam Turbine Historical.jmp.
2. Select **Analyze > Quality and Process > Control Chart > Multivariate Control Chart.**
3. Select all of the columns and click **Y, Columns**.
4. Click **OK**.

Figure 6.2 Initial Multivariate Control Chart

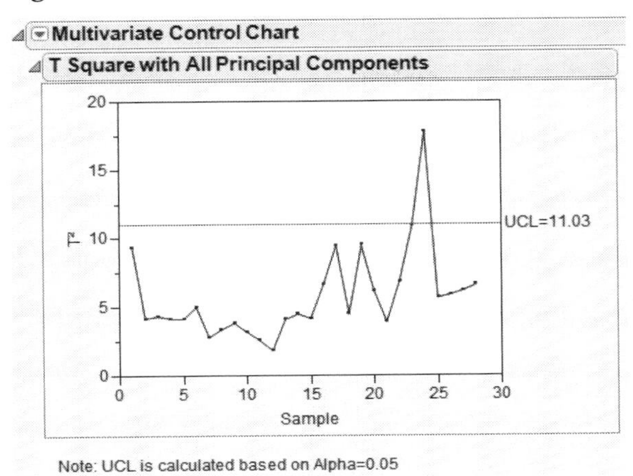

The process seems to be in reasonable statistical control, because there is only one out-of-control point. Therefore, it is appropriate to create targets based on this data.

Step 2: Save Target Statistics

1. From the red triangle menu, select **Save Target Statistics.**

   This creates a new data table containing target statistics for the process.

Figure 6.3 Target Statistics for Steam Turbine Data

	Ref_Stats	Fuel	Steam Flow	Steam Temp	MW	Cool Temp	Pressure
1	_SampleSize	28	28	28	28	28	28
2	_NumSample	1	1	1	1	1	1
3	_Mean	237595.78571	179015.78571	846.39285714	20.647142857	53.871428571	29.139285714
4	_Std	7247.6859825	4374.3063819	2.9481857034	0.5341650261	0.2088010623	0.0497347461
5	_Corr_Fuel	1	0.8714382899	-0.549875041	0.8558570808	-0.270049819	-0.469928462
6	_Corr_Steam Flow	0.8714382899	1	-0.629023927	0.9852529223	-0.223127002	-0.533056185
7	_Corr_Steam Temp	-0.549875041	-0.629023927	1	-0.595214609	0.2475387217	0.2192147319
8	_Corr_MW	0.8558570808	0.9852529223	-0.595214609	1	-0.207305813	-0.50447312
9	_Corr_Cool Temp	-0.270049819	-0.223127002	0.2475387217	-0.207305813	1	0.3617461646
10	_Corr_Pressure	-0.469928462	-0.533056185	0.2192147319	-0.50447312	0.3617461646	1

2. Save the new data table as Steam Turbine Targets.jmp.

Now that you have established targets, create the multivariate control chart that monitors the process.

Step 3: Monitor the Process

1. Select **Help > Sample Data Library** and open Quality Control/Steam Turbine Current.jmp.

This sample data table contains recent observations from the process.

2. Select **Analyze > Quality and Process > Control Chart > Multivariate Control Chart.**
3. Select all of the columns and click **Y, Columns.**
4. Click **Get Targets.**
5. Open the Steam Turbine Targets.jmp table that you saved.
6. Click **OK.**

The default alpha level is set to 0.05. Change it to 0.001.

7. From the red triangle menu, select **Set Alpha Level > Other.**
8. Type 0.001 and click **OK.**

Figure 6.4 Steam Turbine Control Chart

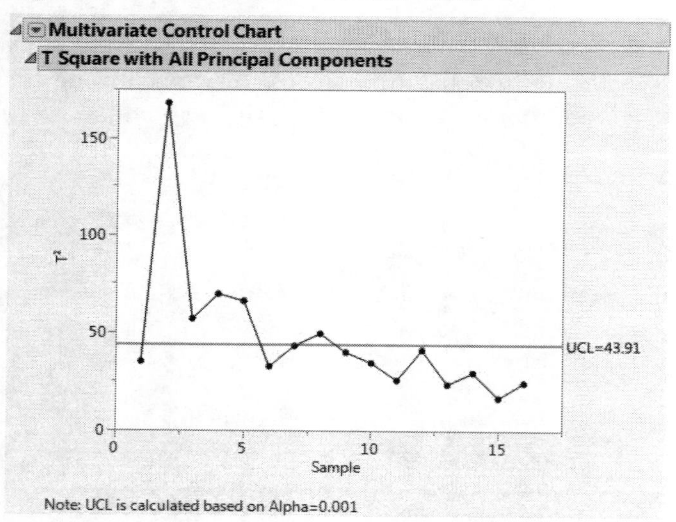

Figure 6.4 shows out-of-control conditions occurring at observations 2, 3, 4, 5, and 8. This result implies that these observations do not conform to the historical data from Steam Turbine Historical.jmp, and that the process should be further investigated. To find an assignable cause, you might want to examine individual univariate control charts or perform another univariate procedure.

Launch the Multivariate Control Chart Platform

Launch the Multivariate Control Chart platform by selecting **Analyze > Quality And Process > Control Chart > Multivariate Control Chart.**

Figure 6.5 The Multivariate Control Chart Launch Window

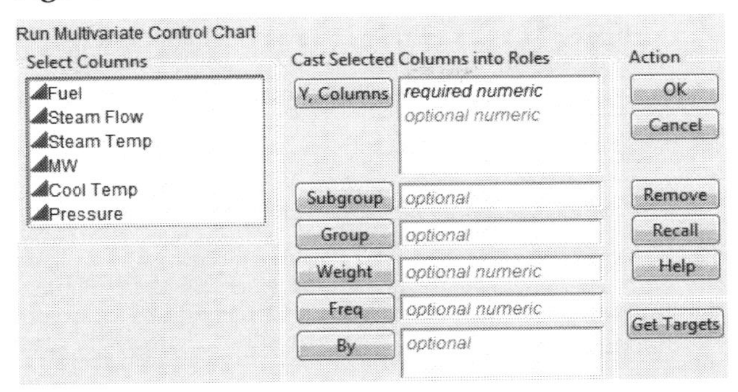

Y, Columns Specify the columns to be analyzed.

Subgroup Enter a column with sub-grouped data. Hierarchically, this group is nested within **Group**.

Group Enter a column that specifies group membership at the highest hierarchical level.

Weight Identifies the data table column whose variables assign weight (such as importance or influence) to the data.

Freq Identifies the data table column whose values assign a frequency to each row. Can be useful when you have summarized data.

By Identifies a column that creates a report consisting of separate analyses for each level of the variable.

Get Targets Click to select a JMP table that contains historical targets for the process.

The Multivariate Control Chart

Use the multivariate control chart to quickly identify shifts in your process and to monitor your process for special cause indications.

Follow the instructions in "Example of a Multivariate Control Chart" on page 128 to produce the results shown in Figure 6.6.

Figure 6.6 Multivariate Control Chart

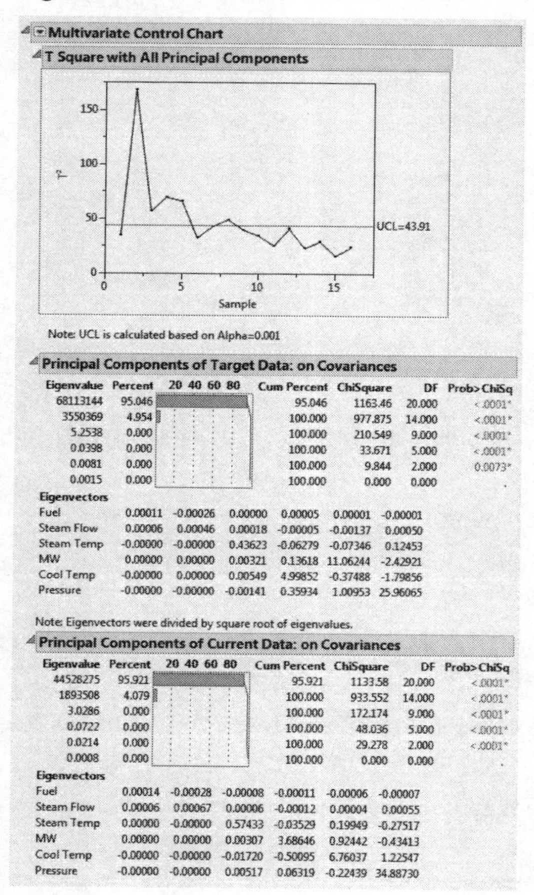

Principal Components of Target Data: on Covariances

Eigenvalue	Percent	Cum Percent	ChiSquare	DF	Prob>ChiSq
68113144	95.046	95.046	1163.46	20.000	<.0001*
3550369	4.954	100.000	977.875	14.000	<.0001*
5.2538	0.000	100.000	210.549	9.000	<.0001*
0.0398	0.000	100.000	33.671	5.000	<.0001*
0.0081	0.000	100.000	9.844	2.000	0.0073*
0.0015	0.000	100.000	0.000	0.000	.

Eigenvectors

Fuel	0.00011	-0.00026	0.00000	0.00005	0.00001	-0.00001
Steam Flow	0.00006	0.00046	0.00018	-0.00005	-0.00137	0.00050
Steam Temp	-0.00000	-0.00000	0.43623	-0.06279	-0.07346	0.12453
MW	0.00000	0.00000	0.00321	0.13618	11.06244	-2.42921
Cool Temp	-0.00000	0.00000	0.00549	4.99852	-0.37488	-1.79856
Pressure	-0.00000	-0.00000	-0.00141	0.35934	1.00953	25.96065

Note: Eigenvectors were divided by square root of eigenvalues.

Principal Components of Current Data: on Covariances

Eigenvalue	Percent	Cum Percent	ChiSquare	DF	Prob>ChiSq
44528275	95.921	95.921	1133.58	20.000	<.0001*
1893508	4.079	100.000	933.552	14.000	<.0001*
3.0286	0.000	100.000	172.174	9.000	<.0001*
0.0722	0.000	100.000	48.036	5.000	<.0001*
0.0214	0.000	100.000	29.278	2.000	<.0001*
0.0008	0.000	100.000	0.000	0.000	.

Eigenvectors

Fuel	0.00014	-0.00028	-0.00008	-0.00011	-0.00006	-0.00007
Steam Flow	0.00006	0.00067	0.00006	-0.00012	0.00004	0.00055
Steam Temp	0.00000	-0.00000	0.57433	-0.03529	0.19949	-0.27517
MW	0.00000	0.00000	0.00307	3.68646	0.92442	-0.43413
Cool Temp	-0.00000	-0.00000	-0.01720	-0.50095	6.76037	1.22547
Pressure	-0.00000	-0.00000	0.00517	0.06319	-0.22439	34.88730

Tip: For information about additional options, see "Multivariate Control Chart Platform Options" on page 133.

The multivariate control chart plots Hotelling's T^2 statistic. The calculation for the control limit differs based on whether targets have been specified. To understand how the T^2 statistic and the UCL (Upper Control Limit) are calculated, see "Statistical Details for Multivariate Control Charts" on page 142. For more details about control limits, see Tracy, et al., 1992.

In this example, the Principal Components reports for both data sets indicate that the first eigenvalue, corresponding to the first principal component, explains about 95% of the total variation in the variables. The values in both Eigenvectors tables indicate that the first principal component is driven primarily by the variables Fuel and Steam Flow. You can use this information to construct a potentially more sensitive control chart based only on this first component. For more details about the Principal Components reports, see "Principal Components" on page 135.

Multivariate Control Chart Platform Options

The following options are available from the platform red triangle menu:

T Square Chart Shows the T^2 chart. Hotelling's T^2 chart is a multivariate extension of the X-bar chart that takes correlation into account.

T Square Partitioned Constructs multivariate control charts based on the principal components of Y. Specify the number of major principal components for T^2. See "T Square Partitioned" on page 134.

Set Alpha Level Set the α level used to calculate the control limit. The default is α=0.05.

Show Covariance Shows the covariance. Covariance is a measure of the linear relationship between two variables.

Show Correlation Shows the correlation report.

Show Inverse Covariance Shows the inverse, or if it is singular, a generalized inverse of the covariance matrix.

Show Inverse Correlation Shows the inverse, or if it is singular, a generalized inverse of the correlation matrix.

Show Means Shows the means for each group.

Save T Square Creates a new column in the data table containing T^2 values.

Save T Square Formula Creates a new column in the data table. Stores a formula in the column that calculates the T^2 values.

Save Target Statistics Creates a new data table containing target statistics for the process. Target statistics include: sample size, the number of samples, mean, standard deviation, and any correlations.

Change Point Detection (Not available for sub-grouped data) Shows a Change Point Detection plot of test statistics by row number and indicates the row number where the change point appears. See "Change Point Detection" on page 134.

Principal Components Shows reports showing eigenvalues and their corresponding eigenvectors. Principal components help you understand which of the many variables you might be monitoring are primarily responsible for the variation in your process. See "Principal Components" on page 135.

Save Principal Components Creates new columns in the data table that contain the scaled principal components.

T Square Partitioned

If you are monitoring a large number of correlated process characteristics, you can use the T Square Partitioned option to construct a control chart based on principal components. If a small number of principal components explains a large portion of the variation in your measurements, then a multivariate control chart based on these big components might be more sensitive than a chart based on your original, higher-dimensional data.

The T Square Partitioned option is also useful when your covariance matrix is ill-conditioned. When this is the case, components with small eigenvalues, explaining very little variation, can have a large, and misleading, impact on T^2. It is useful to separate out these less important components when studying process behavior.

Once you select the T Square Partitioned option, you need to decide how many major principal components to use.

The option creates two multivariate control charts: T Square with Big Principal Components and T Square with Small Principal Components. Suppose that you enter *r* as the number of major components when you first select the option. The chart with Big Principal Components is based on the *r* principal components corresponding to the *r* largest eigenvalues. These are the *r* components that explain the largest amount of variation, as shown in the Percent and Cum Percent columns in the Principal Components: on Covariances reports. The chart with Small Principal Components is based on the remaining principal components.

For a given subgroup, its T2 value in the Big Principal Components chart and its T^2 value in the Small Principal Components chart sum to its overall T^2 statistic presented in the T^2 with All Principal Components report. For details about how the partitioned T^2 values are calculated, see Kourti, T. and MacGregor, J. F., 1996.

Change Point Detection

When the data set consists of multivariate individual observations, a control chart can be developed to detect a shift in the mean vector, the covariance matrix, or both. This method partitions the data and calculates likelihood ratio test statistics for a shift. The statistic that is plotted on the control chart is an observation's likelihood ratio test statistic divided by the product of:

- Its approximate expected value assuming no shift.
- An approximate value for an upper control limit.

Division by the approximate upper control limit allows the points to be plotted against an effective upper control limit of 1. A Change Point Detection plot readily shows the change point for a shift occurring at the maximized value of the control chart statistic. The Change Point Detection implementation in JMP is based on Sullivan and Woodall (2000) and is described in “Statistical Details for Change Point Detection” on page 145.

Note: The Change Point Detection method is designed to show a single shift in the data. Detect multiple shifts by recursive application of this method.

Note the following about the Change Point Detection plot:

- Values above 1.0 indicate a possible shift in the data.
- Control chart statistics for the Change Point Detection plot are obtained by dividing the likelihood ratio statistic of interest (either a mean vector or a covariance matrix) by a normalizing factor.
- The change point of the data occurs for the observation having the maximum test statistic value for the Change Point Detection plot.

Note the following about the scatterplot matrix:

- This plot shows the shift in the sample mean vector.
- In the "Example of Change Point Detection" on page 141, data are divided into two groups. The first 24 observations are classified as the first group. The remaining observations are classified as the second group.

Principal Components

The Principal Components report contain the following information:

Eigenvalue Eigenvalues for the covariance matrix.

Percent Percent variation explained by the corresponding eigenvector. Also shows an accompanying bar chart.

Cum Percent Cumulative percent variation explained by eigenvectors corresponding to the eigenvalues.

ChiSquare Provides a test of whether the correlation remaining in the data is of a random nature. This is a Bartlett test of sphericity. When this test rejects the null hypothesis, this implies that there is structure remaining in the data that is associated with this eigenvalue.

DF Degrees of freedom associated with the Chi-square test.

Prob > ChiSq *p*-value for the test.

Eigenvectors Table of eigenvectors corresponding to the eigenvalues. Note that each eigenvector is divided by the square root of its corresponding eigenvalue.

For more information about principal components, see the Principal Components chapter in the *Multivariate Methods* book.

Additional Examples of Multivariate Control Charts

This section contains additional examples of creating multivariate control charts to monitor an individual process and detect change points.

Example of Monitoring a Process Using Sub-Grouped Data

The workflow for monitoring a multivariate process with sub-grouped data is similar to the one for individual data. See "Example of a Multivariate Control Chart" on page 128. You create an initial control chart to save target statistics and then use these targets to monitor the process.

Step 1: Determine Whether the Process Is Stable

1. Select **Help > Sample Data Library** and open Quality Control/Aluminum Pins Historical.jmp.
2. Select **Analyze > Quality and Process > Control Chart > Multivariate Control Chart.**
3. Select all of the Diameter and Length columns and click **Y, Columns.**
4. Select subgroup and click **Subgroup.**
5. Click **OK.**

Figure 6.7 Multivariate Control Chart for Sub-Grouped Data, Step 1

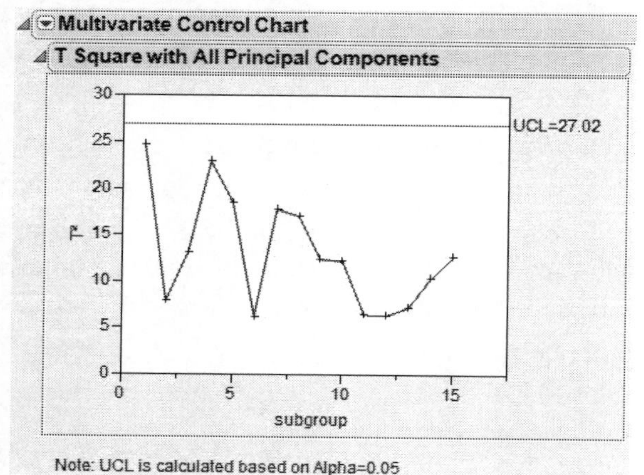

The process appears to be in statistical control, making it appropriate to create targets using this data.

Step 2: Save Target Statistics

1. From the red triangle menu, select **Save Target Statistics.**

This creates a new data table containing target statistics for the process.

2. Save the new data table as Aluminum Pins Targets.jmp.

   Now that you have established targets, create the multivariate control chart that monitors the process.

Step 3: Monitor the Process

1. Select **Help > Sample Data Library** and open Quality Control/Aluminum Pins Current.jmp.

   This sample data table contains recent observations from the process.

2. Select **Analyze > Quality and Process > Control Chart > Multivariate Control Chart**.
3. Select all of the Diameter and Length columns and click **Y, Columns**.
4. Select subgroup and click **Subgroup**.
5. Click **Get Targets**.
6. Open the Aluminum Pins Targets.jmp table that you saved.
7. Click **OK**.
8. From the red triangle menu, select **Show Means**.

   The Show Means option gives the means for each subgroup. You can then observe which groups are most dissimilar from each other.

Figure 6.8 Multivariate Control Chart for Sub-Grouped Data, Step 3

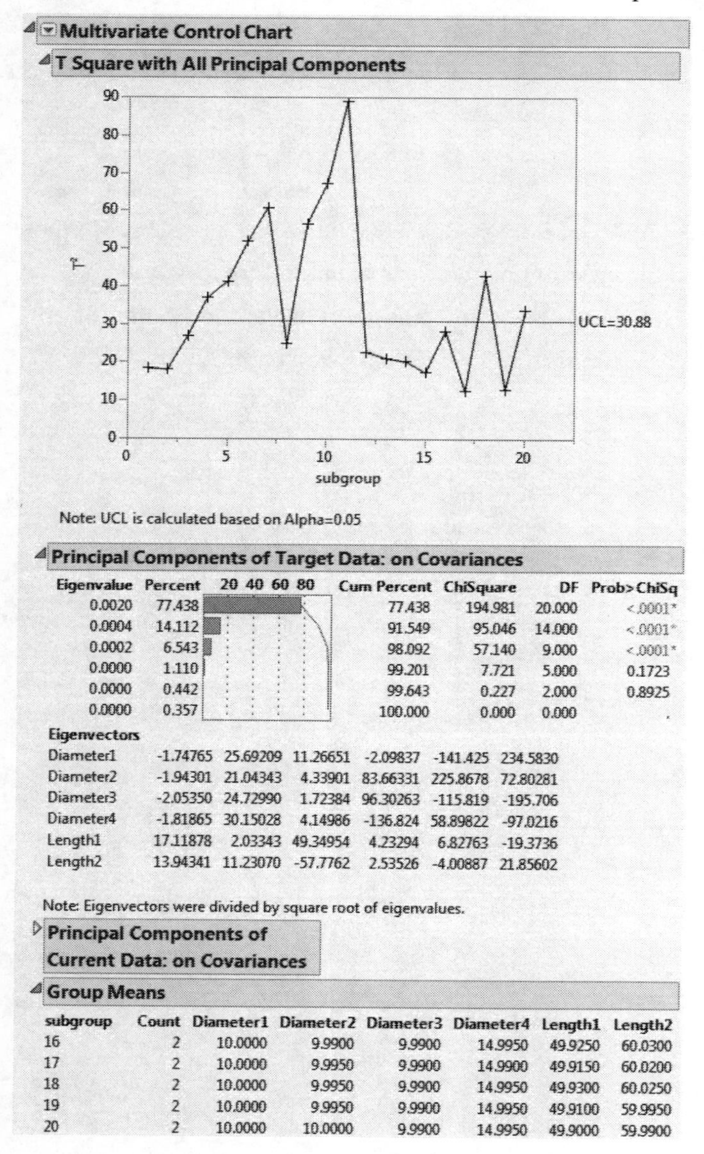

Principal Components of Target Data: on Covariances

Eigenvalue	Percent	Cum Percent	ChiSquare	DF	Prob>ChiSq
0.0020	77.438	77.438	194.981	20.000	<.0001*
0.0004	14.112	91.549	95.046	14.000	<.0001*
0.0002	6.543	98.092	57.140	9.000	<.0001*
0.0000	1.110	99.201	7.721	5.000	0.1723
0.0000	0.442	99.643	0.227	2.000	0.8925
0.0000	0.357	100.000	0.000	0.000	.

Eigenvectors

Diameter1	-1.74765	25.69209	11.26651	-2.09837	-141.425	234.5830
Diameter2	-1.94301	21.04343	4.33901	83.66331	225.8678	72.80281
Diameter3	-2.05350	24.72990	1.72384	96.30263	-115.819	-195.706
Diameter4	-1.81865	30.15028	4.14986	-136.824	58.89822	-97.0216
Length1	17.11878	2.03343	49.34954	4.23294	6.82763	-19.3736
Length2	13.94341	11.23070	-57.7762	2.53526	-4.00887	21.85602

Note: Eigenvectors were divided by square root of eigenvalues.

Principal Components of Current Data: on Covariances

Group Means

subgroup	Count	Diameter1	Diameter2	Diameter3	Diameter4	Length1	Length2
16	2	10.0000	9.9900	9.9900	14.9950	49.9250	60.0300
17	2	10.0000	9.9950	9.9900	14.9900	49.9150	60.0200
18	2	10.0000	9.9950	9.9900	14.9950	49.9300	60.0250
19	2	10.0000	9.9950	9.9900	14.9950	49.9100	59.9950
20	2	10.0000	10.0000	9.9900	14.9950	49.9000	59.9900

Figure 6.8 shows indications of instability at subgroups 4-7, 9-11, 18, and 20. This result implies that these observations do not conform to the historical data from Aluminum Pins Historical.jmp, and that the process should be further investigated. To determine why the process was out of control at these points, you might want to examine individual univariate control charts or perform another univariate procedure.

An alternative method to monitoring this process is based on the big principal components. In this example, for the historical data, the first three principal components account for about 98% of the variation. Based on this, you might construct a chart for the first three principal components. Then you would monitor current data using those three components. The control limits for the chart used in monitoring the process should be based on the corresponding chart for the historical data.

Example of T Square Partitioned

Use T Square Partitioned to separate out the more important components from the less important components when studying process behavior. In this example, the coating on each of 50 bars was measured at 12 uniformly spaced locations across the bar. You want to examine the variation in the measurements and determine whether the causes of variation need to be investigated further.

1. Select **Help > Sample Data Library** and open Quality Control/Thickness.jmp.
2. Select **Analyze > Quality and Process > Control Chart > Multivariate Control Chart.**
3. Select all of the Thickness columns and click **Y, Columns.**
4. Click **OK.**

   The current alpha level is set to 0.05, which corresponds to a 5% false alarm rate. You want to set the false alarm rate to 1%.
5. Change the alpha level by selecting **Set Alpha Level** and choosing **0.01** from the red triangle menu.

Figure 6.9 Initial Multivariate Control Chart for Thickness.jmp

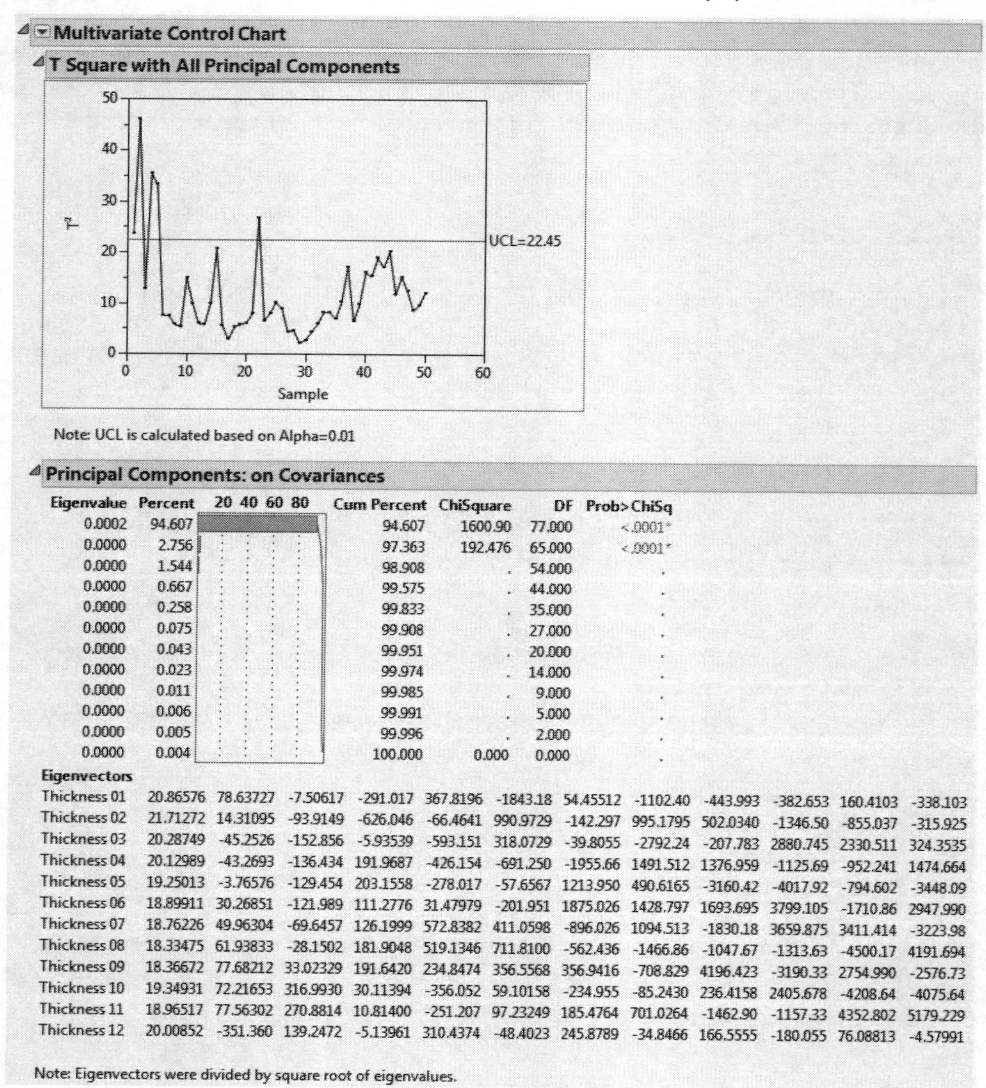

Eigenvalue	Percent	20 40 60 80	Cum Percent	ChiSquare	DF	Prob>ChiSq
0.0002	94.607		94.607	1600.90	77.000	<.0001*
0.0000	2.756		97.363	192.476	65.000	<.0001*
0.0000	1.544		98.908	.	54.000	.
0.0000	0.667		99.575	.	44.000	.
0.0000	0.258		99.833	.	35.000	.
0.0000	0.075		99.908	.	27.000	.
0.0000	0.043		99.951	.	20.000	.
0.0000	0.023		99.974	.	14.000	.
0.0000	0.011		99.985	.	9.000	.
0.0000	0.006		99.991	.	5.000	.
0.0000	0.005		99.996	.	2.000	.
0.0000	0.004		100.000	0.000	0.000	.

Eigenvectors

Thickness 01	20.86576	78.63727	-7.50617	-291.017	367.8196	-1843.18	54.45512	-1102.40	-443.993	-382.653	160.4103	-338.103
Thickness 02	21.71272	14.31095	-93.9149	-626.046	-66.4641	990.9729	-142.297	995.1795	502.0340	-1346.50	-855.037	-315.925
Thickness 03	20.28749	-45.2526	-152.856	-5.93539	-593.151	318.0729	-39.8055	-2792.24	-207.783	2880.745	2330.511	324.3535
Thickness 04	20.12989	-43.2693	-136.434	191.9687	-426.154	-691.250	-1955.66	1491.512	1376.959	-1125.69	-952.241	1474.664
Thickness 05	19.25013	-3.76576	-129.454	203.1558	-278.017	-57.6567	1213.950	490.6165	-3160.42	-4017.92	-794.602	-3448.09
Thickness 06	18.89911	30.26851	-121.989	111.2776	31.47979	-201.951	1875.026	1428.797	1693.695	3799.105	-1710.86	2947.990
Thickness 07	18.76226	49.96304	-69.6457	126.1999	572.8382	411.0598	-896.026	1094.513	-1830.18	3659.875	3411.414	-3223.98
Thickness 08	18.33475	61.93833	-28.1502	181.9048	519.1346	711.8100	-562.436	-1466.86	-1047.67	-1313.63	-4500.17	4191.694
Thickness 09	18.36672	77.68212	33.02329	191.6420	234.8474	356.5568	356.9416	-708.829	4196.423	-3190.33	2754.990	-2576.73
Thickness 10	19.34931	72.21653	316.9930	30.11394	-356.052	59.10158	-234.955	-85.2430	236.4158	2405.678	-4208.64	-4075.64
Thickness 11	18.96517	77.56302	270.8814	10.81400	-251.207	97.23249	185.4764	701.0264	-1462.90	-1157.33	4352.802	5179.229
Thickness 12	20.00852	-351.360	139.2472	-5.13961	310.4374	-48.4023	245.8789	-34.8466	166.5555	-180.055	76.08813	-4.57991

Note: Eigenvectors were divided by square root of eigenvalues.

The overall control chart in Figure 6.9 suggests that special causes have affected bars 1, 2, 4, 5, and 22. Looking at the Principal Components report, you can see that almost 95% of the variation in the 12 thickness measurements is explained by the first principal component. You want to study the variation associated with this principal component further.

6. From the red triangle menu, select **T Square Partitioned**.
7. Accept the default value of 1 principal component by clicking **OK**.

Figure 6.10 T Square Partitioned Control Charts

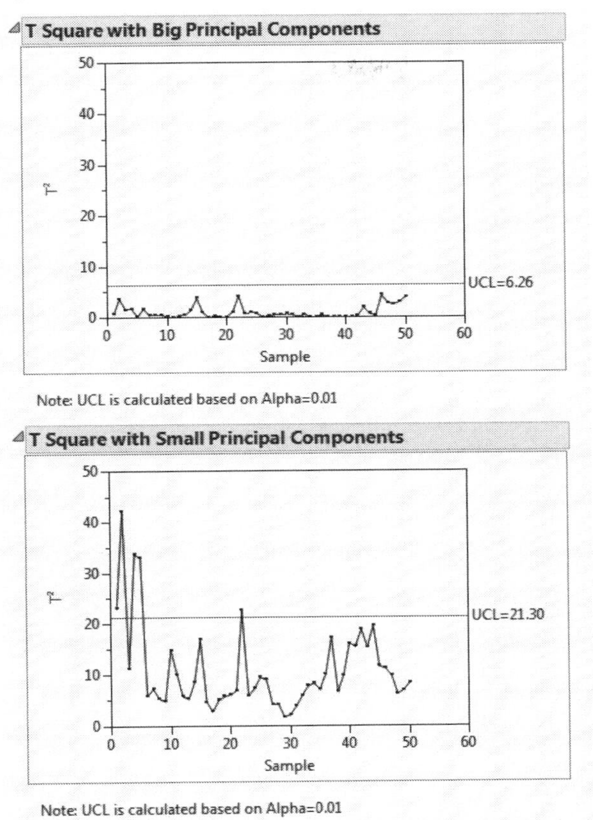

In contrast to the Principal Components report, the T Square with Big Principal Components chart, which reflects variation for only the first component, shows no evidence of special causes. The T Square with Small Principal Components chart shows that the special cause indications reside in the remaining smaller components. These smaller components do not explain much variation, and likely represent random noise. Therefore, you might conclude that the variation in the thickness measurements is not a major cause for concern.

Example of Change Point Detection

Use change point detection to find the point at which a shift occurs in your data.

1. Select **Help > Sample Data Library** and open Quality Control/Gravel.jmp.
2. Select **Analyze > Quality and Process > Control Chart > Multivariate Control Chart**.
3. Select Large and Medium and click **Y, Columns**.
4. Click **OK**.

5. From the red triangle menu, select **Change Point Detection.**

Figure 6.11 Change Point Detection for Gravel.jmp

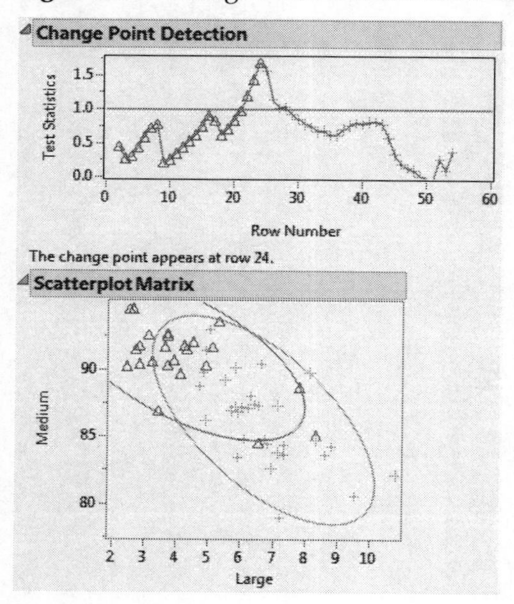

Tip: You might need to drag the axes to see the density ellipses for the two groups, depending on your data.

In the Change Point Detection plot, values above 1.0 indicate a possible shift in the data. At least one shift is apparent; the change point occurs at observation 24 and the shift occurs immediately after observation 24. The 95% prediction regions for the two groups have approximately the same size, shape, and orientation, visually indicating that the sample covariance matrices are similar.

Statistical Details for Multivariate Control Charts

This section contains statistical details for individual data, sub-grouped data, additivity, and change point detection.

Statistical Details for Individual Data

Consider measurements that are not sub-grouped, that is, where the natural subgroup size is one. Denote the number of samples by n and the number of characteristics measured by p. Then the T^2 statistic is defined as follows:

$$T^2 = (X_i - \overline{X})'S^{-1}(X_i - \overline{X})$$

where:

X_i is the column vector of p measurements for the ith sample

$\overline{X}$ is the sample mean of the X_i

S is the sample covariance matrix for the X_i

These are the points plotted on the multivariate control chart.

When computing control limits based on historical data (Step 1), the UCL is based on the beta distribution. Specifically, the UCL is defined as follows:

$$UCL = \frac{(n-1)^2}{n}\beta\left(1-\alpha, \frac{p}{2}, \frac{n-p-1}{2}\right)$$

where:

p is number of variables

n is the sample size

Once you have specified targets (Step 2), new observations are independent of the historical statistics. In this case, the UCL is a function of the F-distribution and partially depends on the number of observations in the data from which the targets are calculated.

If n is less than or equal to 100, the UCL is defined as:

$$UCL = \frac{p(n+1)(n-1)}{n(n-p)}F(1-\alpha, p, n-p)$$

If n is greater than 100, the UCL is defined as:

$$UCL = \frac{p(n-1)}{n-p}F(1-\alpha, p, n-p)$$

In both equations,

p is number of variables

n is the sample size for the data from which the targets were calculated

Statistical Details for Sub-Grouped Data

Consider the case where p characteristics are monitored and where m subgroups of size n are obtained. A value of the T^2 statistic is plotted for each subgroup. The T^2 statistic for the ith subgroup is defined as follows:

$$T^2 = (\overline{X}_i - \overline{\overline{X}})'\overline{S}^{-1}(\overline{X}_i - \overline{\overline{X}})$$

where:

$\overline{X}_i$ is the mean of the n column vectors of p measurements for the ith sample

$\overline{X} = \frac{1}{m}\sum_{i=1}^{m} \overline{X}_i$ is the overall mean of the observations

S_i is the sample covariance matrix for the n observations in the ith subgroup

$\overline{S} = \frac{1}{m}\sum_{i=1}^{m} S_i$ is the mean of the within-subgroup covariance matrices

Using historical data (Step 1), before specifying targets, the UCL is defined as follows:

$$UCL = \frac{p(m-1)(n-1)}{mn-m-p+1}F(1-\alpha, p, mn-m-p+1)$$

where:

p is number of characteristics
n is the sample size for each subgroup
m is the number of subgroups

When you compare current data to historical targets, the UCL is defined as follows:

$$UCL = \frac{p(m+1)(n-1)}{mn-m-p+1}F(1-\alpha, p, mn-m-p+1)$$

where:

p is number of variables
n is the sample size for each subgroup
m is the number of subgroups

Statistical Details for Additivity

When a sample of mn independent normal observations is grouped into m rational subgroups of size n, the distance between the mean $\overline{Y}_j$ of the jth subgroup and the expected value μ is T^2_M. Note that the components of the T^2 statistic are additive, much like sums of squares. In other words, the following computation is true:

$$T^2_A = T^2_M + T^2_D$$

Let T^2_M represent the distance from a target value:

$$T^2_M = n(\overline{Y}_j - \mu)'S_P^{-1}(\overline{Y}_j - \mu)$$

The internal variability is defined as follows:

$$T_D^2 = \sum_{j=1}^{n} (Y_j - \bar{Y})' S_P^{-1} (Y_j - \bar{Y})$$

The overall variability is defined as follows:

$$T_A^2 = \sum_{j=1}^{n} (\bar{Y}_j - \mu)' S_P^{-1} (\bar{Y}_j - \mu)$$

Statistical Details for Change Point Detection

This discussion follows the development in Sullivan and Woodall, 2000.

Assumptions

Denote a multivariate distribution of dimension p with mean vector μ_i and covariance matrix Σ_i by $N_p(\mu_i,\Sigma_i)$. Suppose that the x_i are m (where $m > p$) independent observations from such a distribution:

$$x_i \sim N_p(\mu_i,\Sigma_i), \quad i = 1, \ldots, m$$

If the process is stable, the means μ_i and the covariance matrices Σ_i equal a common value so that the x_i have a $N_p(\mu, \Sigma)$ distribution.

Suppose that a single change occurs in either the mean vector or the covariance matrix, or both, between the m_1 and m_1+1 observations. Then the following conditions hold:

- Observations *1* through m_1 have the same mean vector and the same covariance matrix (μ_a,Σ_a).
- Observations $m_1 + 1$ to m have the same mean vector and covariance matrix (μ_b,Σ_b).
- One of the following occurs:
  - If the change affects the mean, $\mu_a \neq \mu_b$.
  - If the change affects the covariance matrix, $\Sigma_a \neq \Sigma_b$.
  - If the change affects both the mean and the covariance matrix, $\mu_a \neq \mu_b$ and $\Sigma_a \neq \Sigma_b$.

Overview

A likelihood ratio test approach is used to identify changes in one or both of the mean vector and covariance matrix. The likelihood ratio test statistic is used to compute a control chart statistic that has an approximate upper control limit of 1. The control chart statistic is plotted for all possible m_1 values. If any observation's control chart statistic exceeds the upper control limit of 1, this is an indication that a shift occurred. Assuming that exactly one shift occurs, that shift is considered to begin immediately after the observation with the maximum control chart statistic value.

Likelihood Ratio Test Statistic

The maximum value of the log-likelihood function for the first m_1 observations is given as follows:

$$l_1 = -m_1 k_1 \log[2\pi] - m_1 \log\left[|S_1|_{k_1}\right] - m_1 k_1$$

The equation for l_1 uses the following notation:

- S_1 is the maximum likelihood estimate of the covariance matrix for the first m_1 observations.
- $k_1 = \text{Min}[p, m_1\text{-}1]$ is the rank of the p x p matrix S_1.
- The notation $|S_1|_{k_1}$ denotes the generalized determinant of the matrix S_1, which is defined as the product of its k_1 positive eigenvalues λ_j:

$$|S_1|_{k_1} = \prod_{j=1}^{k_1} \lambda_j$$

The generalized determinant is equal to the ordinary determinant when S_1 has full rank.

Denote the maximum of the log-likelihood function for the subsequent $m_2 = m - m_1$ observations by l_2, and the maximum of the log-likelihood function for all m observations by l_0. Both l_2 and l_0 are given by expressions similar to that given for l_1.

The likelihood ratio test statistic compares the sum $l_1 + l_2$, which is the log-likelihood that assumes a possible shift at m_1, to the likelihood l_0, which assumes no shift. If l_0 is substantially smaller than $l_1 + l_2$, the process is assumed to be unstable.

Twice the log of the likelihood ratio test statistic for a test of whether a change begins at observation $m_1 + 1$ is given as follows:

$$\begin{aligned} \text{lrt}[m_1] &= (l_1 + l_2 - l_0) \\ &= (m_1(p - k_1) + m_2(p - k_2))(1 + \log(2\pi)) \\ &\quad + m\log[|S|] - m_1 \log\left[|S_1|_{k_1}\right] - m_2 \log\left[|S_2|_{k_2}\right] \end{aligned}$$

The distribution of twice the log of the likelihood ratio test statistic is asymptotic to a chi-square distribution with $p(p+3)/2$ degrees of freedom. Large log-likelihood ratio values indicate that the process is unstable.

The Control Chart Statistic

Simulations indicate that the expected value of $\mathrm{lrt}[m_1]$ varies based on the observation's location in the series, and, in particular, depends on p and m. (See Sullivan and Woodall, 2000.)

Approximating formulas for the expected value of $\mathrm{lrt}[m_1]$ are derived by simulation. To reduce the dependence of the expected value on p, $\mathrm{lrt}[m_1]$ is divided by its asymptotic expected value, $p(p+3)/2$.

The formulas for the approximated expected value of $\mathrm{lrt}[m_1]$ divided by $p(p+3)/2$ are given as follows:

$$\mathrm{ev}[m,p,m_1] = \begin{cases} a_p + m_1 b_p, & \text{if } m_1 < p+1 \\ a_p + m - m_1 b_p, & \text{if } m - m_1 < p+1 \\ \dfrac{m-2p-1}{m_1 - pm - p - m_1}, & \text{otherwise} \end{cases}$$

where

$$a_p = \frac{0.08684(p-14.69)(p-2.036)}{(p-2)}$$

and

$$b_p = \frac{0.1228(p-1.839)}{(p-2)}$$

For $p = 2$, the value of $\mathrm{ev}[m,p,m_1]$ when m_1 or $m_2 = 2$ is 1.3505.

Note: The formulas above are not accurate for $p > 12$ or $m < (2p + 4)$. In such cases, simulation should be used to obtain approximate expected values.

An approximate upper control limit that yields a false out-of-control signal with probability approximately 0.05, assuming that the process is stable, is given as follows:

$$\mathrm{UCL}[m,p] \cong (3.338 - 2.115\log[p] + 0.8819(\log[p])^2 - 0.1382(\log[p])^3)$$
$$+ \quad (0.6389 - 0.3518\log[p] + 0.01784(\log[p])^3)\log[m].$$

Note that this formula depends on m and p.

The control chart statistic is defined to be twice the log of the likelihood ratio test statistic divided by $p(p+3)$, divided by its approximate expected value, and also divided by the approximate value of the control limit. Because of the division by the approximate value of the UCL, the control chart statistic can be plotted against an upper control limit of 1. The approximate control chart statistic is given as follows:

$$\hat{y}[m_1] = \frac{2\text{lrt}[m_1]}{p(p+3)(\text{ev}[m,p,m_1]\text{UCL}[m,p])}$$

Chapter **7**

Measurement Systems Analysis

Evaluate a Continuous Measurement Process Using the EMP Method

The Measurement Systems Analysis (MSA) platform assesses the precision, consistency, and bias of a measurement system. Before you can study the process itself, you need to make sure that you can accurately and precisely measure the process. If most of the variation that you see comes from the measuring process itself, then you are not reliably learning about the process. Use MSA to find out how your measurement system is performing.

This chapter covers the EMP method. The Gauge R&R method is described in the "Variability Gauge Charts" chapter on page 175.

Figure 7.1 Example of a Measurement System Analysis

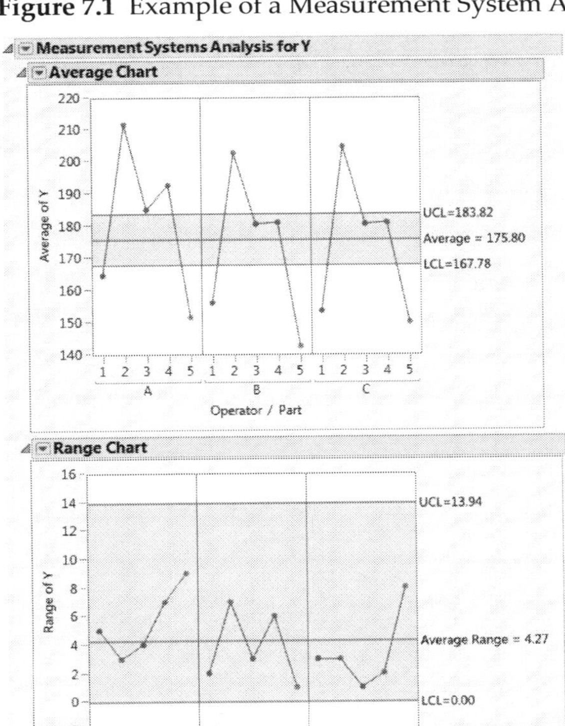

Overview of Measurement Systems Analysis

The EMP (Evaluating the Measurement Process) method in the Measurement Systems Analysis platform is largely based on the methods presented in Donald J. Wheeler's book *EMP III Using Imperfect Data* (2006). The EMP method provides visual information and results that are easy to interpret and helps you improve your measurement system to its full potential.

The Gauge R&R method analyzes how much of the variability is due to operator variation (reproducibility) and measurement variation (repeatability). Gauge R&R is available for many combinations of crossed and nested models, regardless of whether the model is balanced. For more information, see the "Variability Gauge Charts" chapter on page 175.

Within the Six Sigma DMAIC methodology, MSA (Measurement System Analysis) addresses the Measure phase and process behavior charts (or control charts) address the Control phase. MSA helps you predict and characterize future outcomes. You can use the information gleaned from MSA to help you interpret and configure your process behavior charts.

For more information about Control Charts, see the "Control Chart Builder" on page 31.

Example of Measurement Systems Analysis

In this example, three operators measured the same five parts. See how the measurement system is performing, based on how much variation is found in the measurements.

1. Select **Help > Sample Data Library** and open Variability Data/Gasket.jmp.
2. Select **Analyze > Quality and Process > Measurement Systems Analysis.**
3. Assign Y to the **Y, Response** role.
4. Assign Part to the **Part, Sample ID** role.
5. Assign Operator to the **X, Grouping** role.

   Notice that the **MSA Method** is set to **EMP**, the **Chart Dispersion Type** is set to **Range**, and the **Model Type** is set to **Crossed.** See Figure 7.5.
6. Click **OK**.

Figure 7.2 MSA Initial Report

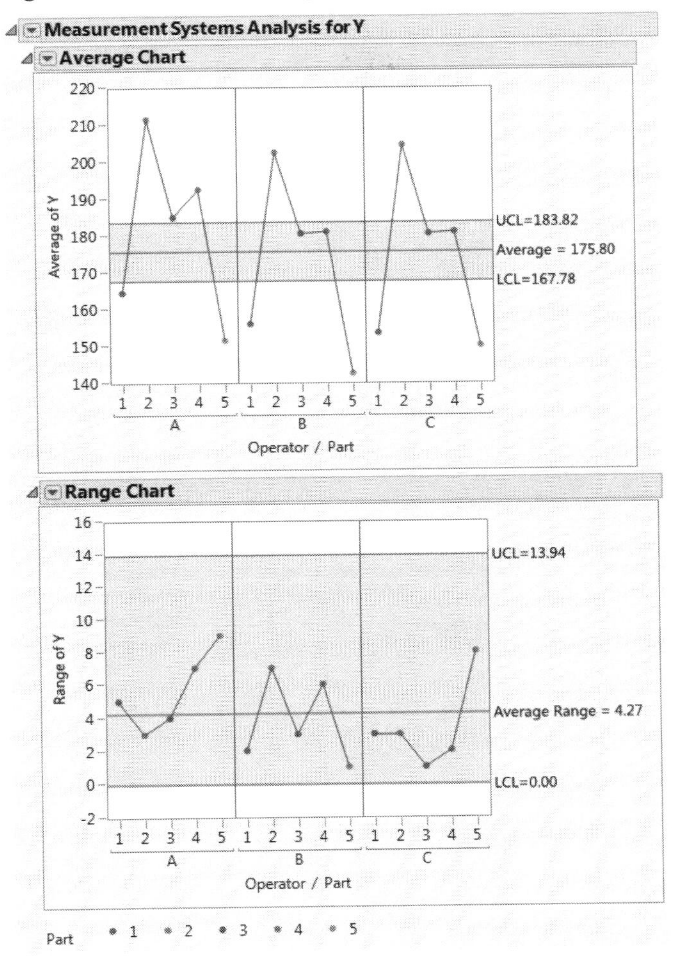

The Average Chart shows the average measurements for each operator and part combination. In this example, the means of the part measurements are generally beyond the control limits. This is a desirable outcome, because it indicates that you can detect part-to-part variation.

The Range Chart shows the variability for each operator and part combination. In this example, the ranges are within the control limits. This is a desirable outcome, because it indicates that the operators are measuring parts in the same way and with similar variation.

The color coding for each part is shown in the legend below the charts.

7. From the red triangle menu next to Measurement Systems Analysis for Y, select **Parallelism Plots**.

Figure 7.3 Parallelism Plot for Operator and Part

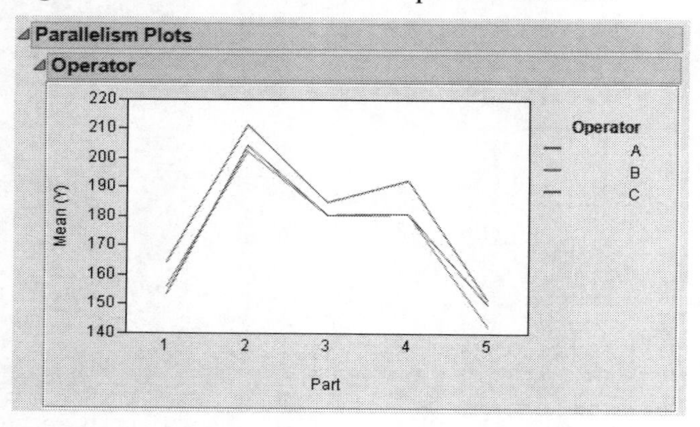

The Parallelism Plots chart shows the average measurements for each part by operator. Because the lines are generally parallel and there is no major crossing, you conclude that there is no interaction between operators and parts.

Tip: Interactions indicate a serious issue that requires further investigation.

---

8. From the red triangle menu next to Measurement Systems Analysis for Y, select **EMP Results**.

Figure 7.4 EMP Results Report

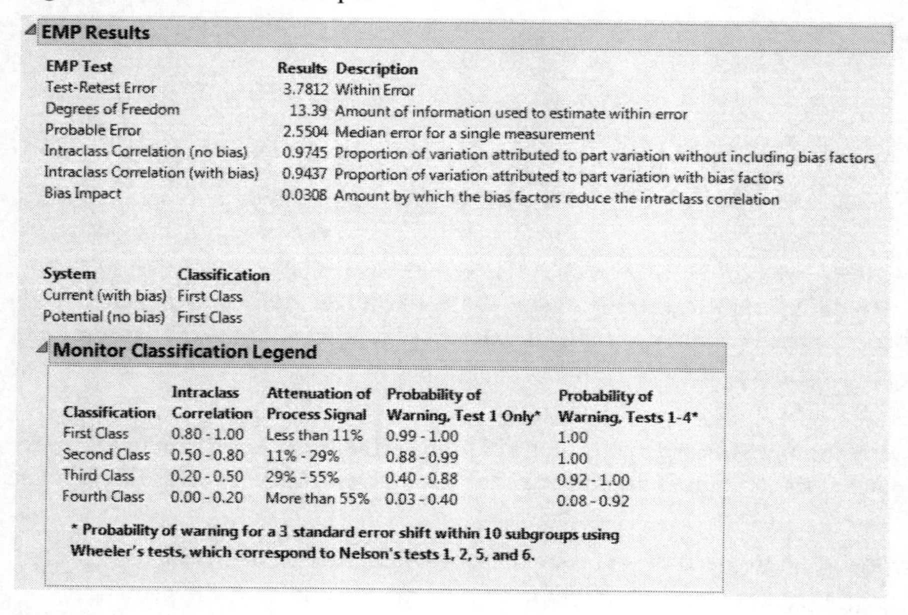

EMP Results

EMP Test	Results	Description
Test-Retest Error	3.7812	Within Error
Degrees of Freedom	13.39	Amount of information used to estimate within error
Probable Error	2.5504	Median error for a single measurement
Intraclass Correlation (no bias)	0.9745	Proportion of variation attributed to part variation without including bias factors
Intraclass Correlation (with bias)	0.9437	Proportion of variation attributed to part variation with bias factors
Bias Impact	0.0308	Amount by which the bias factors reduce the intraclass correlation

System	Classification
Current (with bias)	First Class
Potential (no bias)	First Class

Monitor Classification Legend

Classification	Intraclass Correlation	Attenuation of Process Signal	Probability of Warning, Test 1 Only*	Probability of Warning, Tests 1-4*
First Class	0.80 - 1.00	Less than 11%	0.99 - 1.00	1.00
Second Class	0.50 - 0.80	11% - 29%	0.88 - 0.99	1.00
Third Class	0.20 - 0.50	29% - 55%	0.40 - 0.88	0.92 - 1.00
Fourth Class	0.00 - 0.20	More than 55%	0.03 - 0.40	0.08 - 0.92

* Probability of warning for a 3 standard error shift within 10 subgroups using Wheeler's tests, which correspond to Nelson's tests 1, 2, 5, and 6.

The EMP Results report computes several statistics to help you assess and classify your measurement system. The Intraclass Correlation indicates the proportion of the total variation that you can attribute to the part.

From the EMP Results report, you can conclude the following:

- The Intraclass Correlation values are close to 1, indicating that most of the variation is coming from the part instead of the measurement system.
- The classification is First Class, meaning that the strength of the process signal is weakened by less than 11%.
- There is at least a 99% chance of detecting a warning using Test 1 only.
- There is 100% chance of detecting a warning using Tests 1-4.

Note: For more information about tests and detecting process shifts, see "Shift Detection Profiler" on page 160.

There is no interaction between operators and parts, and there is very little variation in your measurements (the classification is First Class). Therefore, you conclude that the measurement system is performing quite well.

Launch the Measurement Systems Analysis Platform

Launch the Measurement Systems Analysis platform by selecting **Analyze > Quality and Process > Measurement Systems Analysis**.

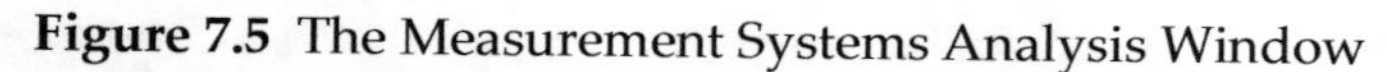

Figure 7.5 The Measurement Systems Analysis Window

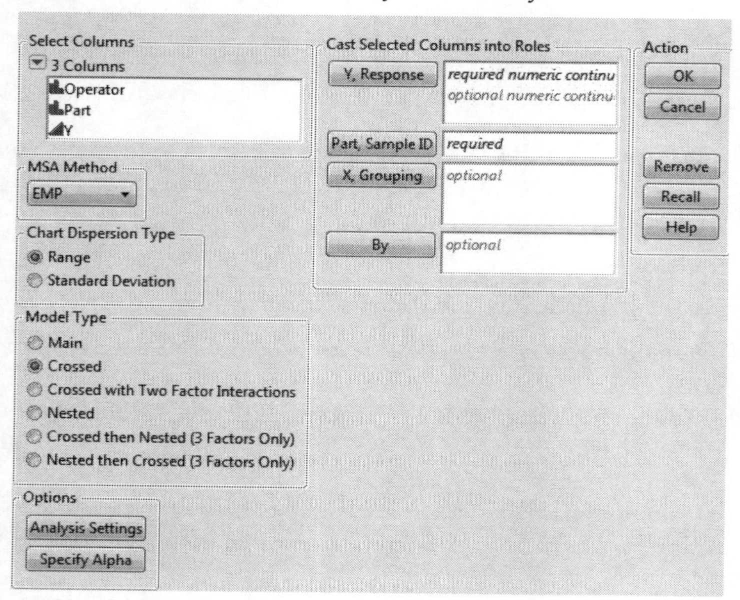

The Measurement Systems Analysis window contains the following features:

Select Columns lists all of the variables in your current data table. Move a selected column into a role.

MSA Method select the method to use: **EMP** (Evaluating the Measurement Process) or **Gauge R&R**. This chapter covers the **EMP** method. For details about the **Gauge R&R** method, see the "Variability Gauge Charts" chapter on page 175.

Chart Dispersion Type designates the type of chart for showing variation. Select the **Range** option or the **Standard Deviation** option.

Note: For the **EMP** method, the chart dispersion type determines how the statistics in the EMP Results report are calculated. If the **Range** option is selected, and you have a one factor or a two factor, balanced, crossed model, the statistics in this report are based on ranges. Otherwise, the statistics in this report are based on standard deviations.

Model Type designates the model type:

- Main: variables with nominal or ordinal modeling types are treated as main effects.
- Crossed: the model is crossed when every level of every factor occurs with every level of every other factor.
- Crossed with Two Factor Interactions: the model is crossed when each level of two factors occurs with every level of the other factor.

– Nested: the model is nested when all levels of a factor appear within only a single level of any other factor.
– Cross then Nested (3 Factors Only): the factors are crossed and then nested for 3 factors.
– Nested then Crossed (3 Factors Only): the factors are nested and then crossed for 3 factors.

Options contains the following options:

– **Analysis Settings** sets the REML maximum iterations and convergence.
– **Specify Alpha** specifies the 1-alpha confidence level.

Y, Response is the column of measurements.

Part, Sample, ID is the column designating the part or unit.

X, Grouping is the column(s) representing grouping variables.

By identifies a column that creates a report consisting of separate analyses for each level of the variable.

Measurement Systems Analysis Platform Options

Platform options appear within the red triangle menu next to Measurement Systems Analysis. Selecting an option creates the respective graph or report in the MSA report window. Deselecting an option removes the graph or report. Choose from the following options:

Average Chart A plot of the average measurement values for each combination of the part and *X* variables. The Average Chart helps you detect product variation despite measurement variation. In an Average Chart, out of control data is desirable because it detects part-to-part variation. See "Average Chart" on page 157.

Range Chart A plot of the variability statistic for each combination of the part and *X* variables. Appears only if you selected **Range** as the Chart Dispersion Type in the launch window. The Range Chart helps you check for consistency within subgroups. In a Range Chart, data within limits is desirable, indicating homogeneity in your error. See "Range Chart or Standard Deviation Chart" on page 157.

Std Dev Chart A plot of the standard deviation statistic for each combination of the part and *X* variables. Appears only if you selected **Standard Deviation** as the Chart Dispersion Type in the launch window. The Standard Deviation Chart helps you check for consistency within subgroups. In a Standard Deviation Chart, data within limits is desirable, indicating homogeneity in your error. See "Range Chart or Standard Deviation Chart" on page 157.

Parallelism Plots An overlay plot that reflects the average measurement values for each part. If the lines are relatively not parallel or crossing, there might be an interaction between the part and *X* variables.

Tip: Interactions indicate a serious issue that requires further investigation. For example, interactions between parts and operators mean that operators are measuring different parts differently, on average. Therefore, measurement variability is not predictable. This issue requires further investigation to find out why the operators do not have the same pattern or profile over the parts.

EMP Results a report that computes several statistics to help you assess and classify your measurement system. See "EMP Results" on page 158.

Effective Resolution a report containing results for the resolution of a measurement system. See "Effective Resolution" on page 159.

Bias Comparison an Analysis of Means chart for testing if the *X* variables have different averages. See "Bias Comparison" on page 165.

Test-Retest Error Comparison an Analysis of Means for Variances or Analysis of Means Ranges chart for testing if any of the groups have different test-retest error levels. See "Test-Retest Error Comparison" on page 165.

Shift Detection Profiler an interactive set of charts that you can adjust to see the probabilities of getting warnings on your process behavior chart. See "Shift Detection Profiler" on page 160.

Variance Components a report containing the estimates of the variance components for the given model. The calculations in this report are based on variances, not ranges. Balanced data uses the EMS method. Unbalanced data uses the REML method.

Note: This report is similar to the Variance Components report in the Variability Chart platform, except that it does not compute Bayesian variance component estimates. For more information, see "Variance Components" on page 183 in the "Variability Gauge Charts" chapter.

EMP Gauge RR Results a report that partitions the variability in the measurements into part variation and measurement system variation. The calculations in this report are based on variances, not ranges.

Note: This report is similar to the Gauge R&R report in the Variability Chart platform, except that the calculation for Reproducibility does not include interactions. For more information about Gauge R&R studies, see "About the Gauge R&R Method" on page 185 in the "Variability Gauge Charts" chapter.

See the JMP Reports chapter in the *Using JMP* book for more information about the following options:

Local Data Filter Shows or hides the local data filter that enables you to filter the data used in a specific report.

Redo Contains options that enable you to repeat or relaunch the analysis. In platforms that support the feature, the Automatic Recalc option immediately reflects the changes that you make to the data table in the corresponding report window.

Save Script Contains options that enable you to save a script that reproduces the report to several destinations.

Save By-Group Script Contains options that enable you to save a script that reproduces the platform report for all levels of a By variable to several destinations. Available only when a By variable is specified in the launch window.

Average Chart

The red triangle menu next to Average Chart contains the following options:

Show Grand Mean draws the overall mean of the *Y* variable on the chart.

Show Connected Means draws lines connecting all of the average measurement values.

Show Control Limits draws lines representing the Upper Control Limit (UCL) and the Lower Control Limit (LCL) and defines those values.

Show Control Limits Shading adds shading between the UCL and LCL.

Show Separators draws vertical lines to delineate between the *X* variables.

Show Data adds the data points to the chart.

Note: You can replace variables in the Average Chart in one of two ways: swap existing variables by dragging and dropping a variable from one axis to the other axis; or, click on a variable in the Columns panel of the associated data table and drag it onto an axis.

Range Chart or Standard Deviation Chart

The red triangle menu next to Range Chart or Standard Deviation Chart contains the following options:

Show Average Dispersion draws the average range or standard deviation on the chart.

Show Connected Points draws lines connecting all of the ranges or standard deviations.

Show Control Limits draws lines representing the Upper Control Limit (UCL) and the Lower Control Limit (LCL) and defines those values.

Show Control Limits Shading adds shading between the UCL and LCL.

Show Separators draws vertical lines to delineate between the *X* variables.

Note: You can replace variables in the Range or Standard Deviation Charts in one of two ways: swap existing variables by dragging and dropping a variable from one axis to the other axis; or, click on a variable in the Columns panel of the associated data table and drag it onto an axis.

EMP Results

Note: The statistics in this report are based on ranges in the following instances: if you selected **EMP** as the **MSA Method** and **Range** as the **Chart Dispersion Type**, and you have a one factor or a two factor, balanced, crossed model. Otherwise, the statistics in this report are based on variances.

The EMP Results report computes several statistics to help you assess and classify your measurement system. Using this report, you can determine the following:

- How your process chart is affected.
- Which tests to set.
- How much the process signal is attenuated.
- How much the bias factors are affecting your system and reducing your potential intraclass correlation coefficient.

The EMP Results report contains the following calculations:

Test-Retest Error indicates measurement variation or repeatability (also known as within error or pure error).

Degrees of Freedom indicates the amount of information used to estimate the within error.

Probable Error the median error for a single measurement. Indicates the resolution quality of your measurement and helps you decide how many digits to use when recording measurements. For more information, see "Effective Resolution" on page 159.

Intraclass Correlation indicates the proportion of the total variation that you can attribute to the part. If you have very little measurement variation, this number is closer to 1.

- **Intraclass Correlation (no bias)** does not take bias or interaction factors into account when calculating the results.
- **Intraclass Correlation (with bias)** takes the bias factors (such as operator, instrument, and so on) into account when calculating the results.
- **Intraclass Correlation (with bias and interaction)** takes the bias and interaction factors into account when calculating the results. This calculation appears only if the model is crossed and uses standard deviation instead of range.

Bias Impact the amount by which the bias factors reduce the Intraclass Correlation.

Bias and Interaction Impact the amount by which the bias and interaction factors reduce the Intraclass Correlation. This calculation appears only if the model is crossed and uses standard deviation instead of range.

Classes of Process Monitors

In order to understand the System and Classification parameters, you must first understand the Monitor Classification Legend.

Figure 7.6 Monitor Classification Legend

Monitor Classification Legend

Classification	Intraclass Correlation	Attenuation of Process Signal	Probability of Warning, Test 1 Only*	Probability of Warning, Tests 1-4*
First Class	0.80 - 1.00	Less than 11%	0.99 - 1.00	1.00
Second Class	0.50 - 0.80	11% - 29%	0.88 - 0.99	1.00
Third Class	0.20 - 0.50	29% - 55%	0.40 - 0.88	0.92 - 1.00
Fourth Class	0.00 - 0.20	More than 55%	0.03 - 0.40	0.08 - 0.92

* Probability of warning for a 3 standard error shift within 10 subgroups using Wheeler's tests, which correspond to Nelson's tests 1, 2, 5, and 6.

This legend describes the following classifications: First, Second, Third, and Fourth Class. Each classification indicates the following:

- the corresponding Intraclass Correlation values
- the amount of process signal attenuation (decrease)
- the chance of detecting a 3 standard error shift within 10 subgroups, using Wheeler's test one or all four tests

Wheeler (2006) identifies four detection tests known as the Western Electric Zone Tests. Within the Shift Detection Profiler, there are eight tests that you can select from. The tests that correspond to the Wheeler tests are the first, second, fifth, and sixth tests.

Tip: To prevent the legend from appearing, deselect **Show Monitor Classification Legend** in the EMP Measurement Systems Analysis platform preferences.

Effective Resolution

The Effective Resolution report helps you determine how well your measurement increments are working. You might find that you need to add or drop digits when recording your measurements, or your current increments might be effective as is. Note the following:

- The Probable Error calculates the median error of a measurement.
- The Current Measurement Increment reflects how many digits you are currently rounding to and is taken from the data as the nearest power of ten. This number is compared to the

Smallest Effective Increment, Lower Bound Increment, and Largest Effective Increment. Based on that comparison, a recommendation is made.

- Large measurement increments have less uncertainty in the last digit, but large median errors. Small measurement increments have small median errors, but more uncertainty in the last digit.

Shift Detection Profiler

Use the Shift Detection Profiler to assess the sensitivity of the control chart that you use to monitor your process. The Shift Detection Profiler estimates the probability of detecting shifts in the product mean or product standard deviation. The control chart limits include sources of measurement error variation. Based on these limits, the Shift Detection Profiler estimates the Probability of Warning. This is the probability that a control chart monitoring the process mean signals a warning over the next *k* subgroups.

You can set the subgroup size that you want to use for your control chart. Note the following:

- If the Subgroup Size equals one, the control chart is an Individual Measurement chart.
- If the Subgroup Size exceeds one, the control chart is an X-bar chart.

You can explore the effect of Subgroup Size on the control chart's sensitivity. You can also explore the benefits of reducing bias and test-retest error.

Figure 7.7 shows the Shift Detection Profiler report for the Gasket.jmp sample data table, found in the Variability Data folder.

Figure 7.7 Shift Detection Profiler for Gasket.jmp

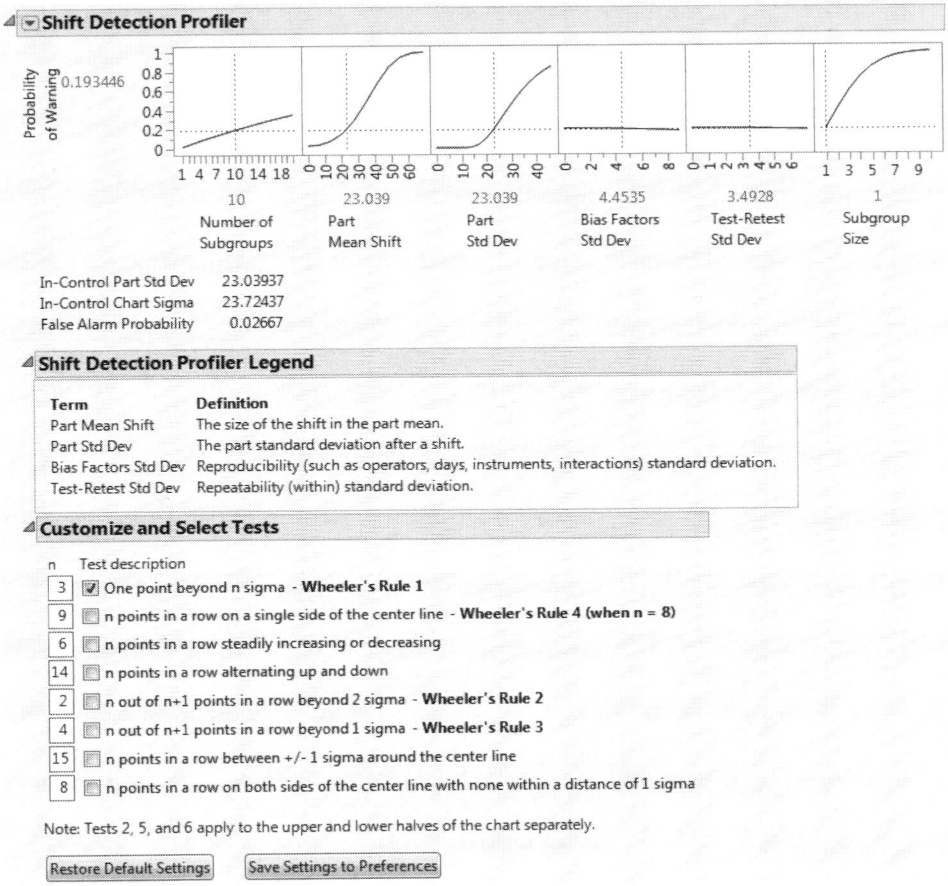

Shift Detection Profiler

Probability of Warning 0.193446

Number of Subgroups	Part Mean Shift	Part Std Dev	Bias Factors Std Dev	Test-Retest Std Dev	Subgroup Size
10	23.039	23.039	4.4535	3.4928	1

In-Control Part Std Dev	23.03937
In-Control Chart Sigma	23.72437
False Alarm Probability	0.02667

Shift Detection Profiler Legend

Term	Definition
Part Mean Shift	The size of the shift in the part mean.
Part Std Dev	The part standard deviation after a shift.
Bias Factors Std Dev	Reproducibility (such as operators, days, instruments, interactions) standard deviation.
Test-Retest Std Dev	Repeatability (within) standard deviation.

Customize and Select Tests

n	Test description
3	One point beyond n sigma - **Wheeler's Rule 1**
9	n points in a row on a single side of the center line - **Wheeler's Rule 4 (when n = 8)**
6	n points in a row steadily increasing or decreasing
14	n points in a row alternating up and down
2	n out of n+1 points in a row beyond 2 sigma - **Wheeler's Rule 2**
4	n out of n+1 points in a row beyond 1 sigma - **Wheeler's Rule 3**
15	n points in a row between +/- 1 sigma around the center line
8	n points in a row on both sides of the center line with none within a distance of 1 sigma

Note: Tests 2, 5, and 6 apply to the upper and lower halves of the chart separately.

Restore Default Settings | Save Settings to Preferences

Probability of Warning

The Probability of Warning is the probability of detecting a change in the process. A change is defined by the Part Mean Shift and the Part Std Dev settings in the Shift Detection Profiler. The probability calculation assumes that the tests selected in the Customize and Select Tests outline are applied to the Number of Subgroups specified in the Profiler.

The control limits for the Individual Measurement chart (Subgroup Size = 1) and the X-bar chart (Subgroup Size > 1) are based on the In-Control Chart Sigma. The In-Control Sigma takes into account the bias factor (reproducibility) variation and the test-retest (repeatability) variation. These are initially set to the values obtained from your MSA study. The In-Control Chart Sigma also incorporates the In-Control Part Std Dev. Both of these values appear beneath the profiler, along with the False Alarm Probability, which is based on the In-Control Chart Sigma.

In-Control Part Std Dev The standard deviation for the true part values, exclusive of measurement errors, for the stable process. The default value for In-Control Part Std Dev is the standard deviation of the part component estimated by the MSA analysis and found in the Variance Components report.

Often, parts for an MSA study are chosen to have specific properties and do not necessarily reflect the part-to-part variation seen in production. For this reason, you can specify the in-control part standard deviation by selecting **Change In-Control Part Std Dev** from the Shift Detection Profiler red triangle menu.

In-Control Chart Sigma The value of sigma used to compute control limits. This value is computed using the In-Control Part Std Dev, the Bias Factors Std Dev, and Test-Retest Std Dev specified in the Shift Detection Profiler, and the Subgroup Size. The reproducibility factors are assumed to be constant within a subgroup.

For a subgroup of size n, control limits are set at the following values:

$$\pm 3(\text{In-Control Chart Sigma})/(\sqrt{n})$$

It follows that the In-Control Chart Sigma is the square root of the sum of the squares of the following terms:

– In-Control Part Std Dev

– Bias Factors Std Dev, as specified in the Shift Detection Profiler, multiplied by $\sqrt{n}$

– Test-Retest Std Dev, as specified in the Shift Detection Profiler

The Bias Factors Std Dev is multiplied by $\sqrt{n}$ to account for the assumption that the reproducibility factors are constant within a subgroup.

JMP updates the In-Control Chart Sigma when you change the In-Control Part Std Dev, the Bias Factors Std Dev, the Test-Retest Std Dev, or the Subgroup Size.

False Alarm Probability The probability that the control chart tests signal a warning when no change in the part mean or standard deviation has occurred. JMP updates the False Alarm Probability when you change the Number of Subgroups or the tests in Customize and Select Tests.

For more information about the Variance Components report, see "Variance Components" on page 183 in the "Variability Gauge Charts" chapter.

Shift Detection Profiler Settings

Number of Subgroups The number of subgroups over which the probability of a warning is computed. If the number of subgroups is set to k, the profiler gives the probability that the control chart signals at least one warning based on these k subgroups. The Number of Subgroups is set to 10 by default. Drag the vertical line in the plot to change the Number of Subgroups.

Part Mean Shift The shift in the part mean. By default, the profiler is set to detect a 1 sigma shift. The initial value is the standard deviation of the part component estimated by the MSA analysis and found in the Variance Components report. Drag the vertical line in the plot or click the value beneath the plot to change the Part Mean Shift.

Part Std Dev The standard deviation for the true part values, exclusive of measurement errors. The initial value for Part Std Dev is the standard deviation of the part component estimated by the MSA analysis and is found in the Variance Components report. Drag the vertical line in the plot or click the value beneath the plot to change the Part Std Dev.

Bias Factors Std Dev The standard deviation of factors related to reproducibility. Bias factors include operator and instrument. The bias factor variation does not include part and repeatability (within) variation. The initial value is derived using the reproducibility and interaction variance components estimated by the MSA analysis and is found in the Variance Components report. Drag the vertical line in the plot or click the value beneath the plot to change the Bias Factors Std Dev.

Test-Retest Std Dev The standard deviation of the test-retest, or repeatability, variation in the model. The initial value is the standard deviation of the Within component estimated by the MSA analysis and is found in the Variance Components report. Drag the vertical line in the plot or click the value beneath the plot to change the Test-Retest Std Dev.

Subgroup Size The sample size used for each subgroup. This is set to 1 by default. You can increase the sample size to investigate improvement in control chart performance. Increasing the sample size from 1 demonstrates what happens when you move from an Individual Measurement chart to an XBar chart. Drag the vertical line in the plot to change the Subgroup Size.

Shift Detection Profiler Options

The red triangle menu for the Shift Detection Profiler provides several options. Only one option is described here.

Change In-Control Part Std Dev Specify a value for the part standard deviation for the stable process. The in-control part standard deviation should reflect the variation of the true part values, exclusive of measurement errors. Enter a new value and click OK.

The In-Control Part Std Dev is originally set to the standard deviation of the part component estimated by the MSA analysis, found in the Variance Components report.

This option is useful if the parts chosen for the EMP study were not a random sample from the process.

Reset Factor Grid Displays a window for each factor allowing you to enter a specific value for the factor's current setting, to lock that setting, and to control aspects of the grid. See the Introduction to Profilers chapter in the *Profilers* book for details.

Factor Settings Submenu that consists of the following options:

Remember Settings Adds an outline node to the report that accumulates the values of the current settings each time the Remember Settings command is invoked. Each remembered setting is preceded by a radio button that is used to reset to those settings.

Copy Settings Script Copies the current Profiler's settings to the clipboard.

Paste Settings Script Pastes the Profiler settings from the clipboard to a Profiler in another report.

Set Script Sets a script that is called each time a factor changes. The set script receives a list of arguments of the form:

```
{factor1 = n1, factor2 = n2, ...}
```

For example, to write this list to the log, first define a function:

```
ProfileCallbackLog = Function({arg},show(arg));
```

Then enter `ProfileCallbackLog` in the Set Script dialog.

Similar functions convert the factor values to global values:

```
ProfileCallbackAssign = Function({arg},evalList(arg));
```

Or access the values one at a time:

```
ProfileCallbackAccess =
Function({arg},f1=arg["factor1"];f2=arg["factor2"]);
```

Shift Detection Profiler Legend

This panel gives a brief description of four of the Shift Detection Profiler settings. For further details, see "Shift Detection Profiler Settings" on page 162.

Tip: To prevent the legend from appearing, deselect **Show Shift Detection Profiler Legend** in the EMP Measurement Systems Analysis platform preferences.

Customize and Select Tests

In the Customize and Select Tests panel, select and customize the tests that you want to apply to the *k* subgroups in your control chart. The eight tests are based on Nelson (1984). For more details about the tests, see "Tests" on page 48 in the "Control Chart Builder" chapter.

The Shift Detection Profiler calculations take these tests into account. The Probability of Warning and False Alarm Probability values increase as you add more tests. Because the calculations are based on a quasi-random simulation, there might be a slight delay as the profiler is updated.

The Customize and Select Tests panel has the following options:

Restore Default Settings If no settings have been saved to preferences, this option resets the selected tests to the first test only. The values of *n* are also reset to the values described in

"Tests" on page 48 in the "Control Chart Builder" chapter. If settings have been saved to preferences, this option resets the selected tests and the values of n to those specified in the preferences.

Note: You can access preferences for control chart tests by selecting **File > Preferences> Platforms > Control Chart Builder**. Custom Tests 1 through 8 correspond to the eight tests shown in Customize and Select Tests.

Save Settings to Preferences Saves the selected tests and the values of n for use in future analyses. These preferences are added to the Control Chart Builder platform preferences.

Bias Comparison

The **Bias Comparison** option creates an Analysis of Means chart. This chart shows the mean values for each level of the grouping variables and compares them with the overall mean. You can use this chart to see whether an operator is measuring parts too high or too low, on average.

The red triangle menu next to Analysis of Means contains the following options:

Set Alpha Level select an option from the most common alpha levels or specify any level using the **Other** selection. Changing the alpha level modifies the upper and lower decision limits.

Show Summary Report shows a report containing group means and decision limits, and reports if the group mean is above the upper decision limit or below the lower decision limit.

Display Options include the following options:

- **Show Decision Limits** draws lines representing the Upper Decision Limit (UDL) and the Lower Decision Limit (LDL) and defines those values.
- **Show Decision Limit Shading** adds shading between the UDL and the LDL.
- **Show Center Line** draws the center line statistic that represents the average.
- **Point Options** changes the chart display to needles, connected points, or points.

Test-Retest Error Comparison

The **Test-Retest Error Comparison** option creates a type of Analysis of Means for Variances or Analysis of Means Ranges chart. This chart shows if there are differences in the test-retest error between operators. For example, you can use this chart to see whether there is an inconsistency in how each operator is measuring. The Analysis of Mean Ranges chart is displayed when ranges are used for variance components.

- For information about the options in the red triangle menu next to Operator Variance Test, see "Bias Comparison" on page 165.
- For more information about Analysis of Means for Variances charts, see "Variance Components" on page 183 in the "Variability Gauge Charts" chapter.

Additional Example of Measurement Systems Analysis

In this example, three operators have measured a single characteristic twice on each of six wafers. Perform a detailed analysis to find out how well the measurement system is performing.

Perform the Initial Analysis

1. Select **Help > Sample Data Library** and open Variability Data/Wafer.jmp.
2. Select **Analyze > Quality and Process > Measurement Systems Analysis.**
3. Assign Y to the **Y, Response** role.
4. Assign Wafer to the **Part, Sample ID** role.
5. Assign Operator to the **X, Grouping** role.

   Notice that the **MSA Method** is set to **EMP**, the **Chart Dispersion Type** is set to **Range**, and the **Model Type** is set to **Crossed.**
6. Click **OK.**

Figure 7.8 Average and Range Charts

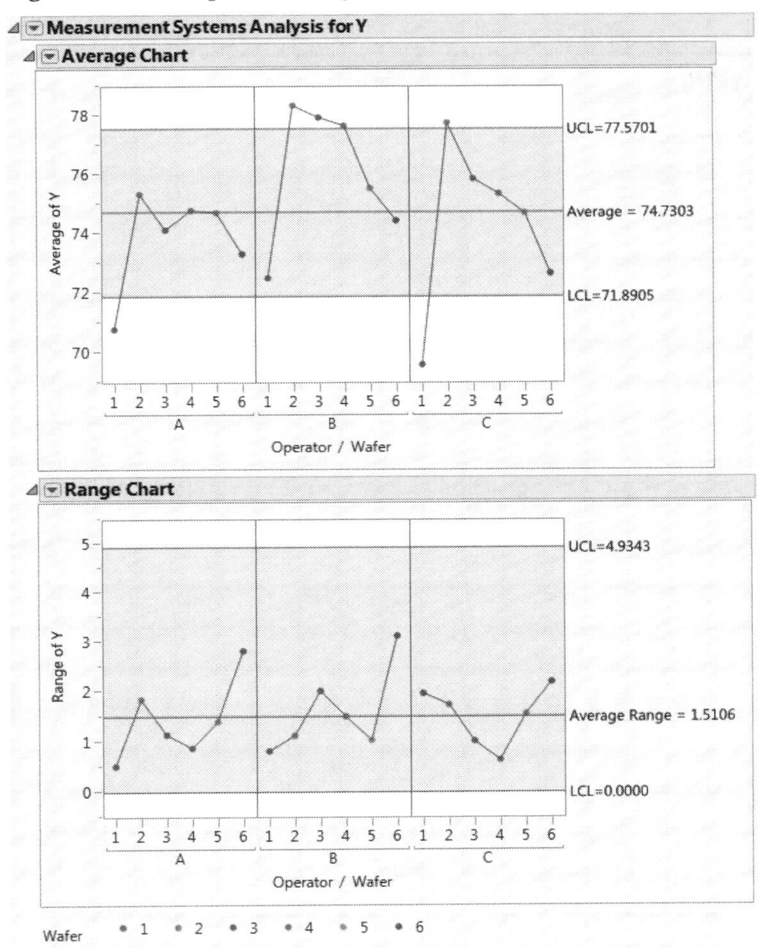

The Average Chart shows that some of the average part measurements fall beyond the control limits. This is desirable, indicating measurable part-to-part variation.

The Range Chart shows no points that fall beyond the control limits. This is desirable, indicating that the operator measurements are consistent within part.

Examine Interactions

Take a closer look for interactions between operators and parts. From the red triangle menu next to Measurement Systems Analysis for Y, select **Parallelism Plots**.

Figure 7.9 Parallelism Plot

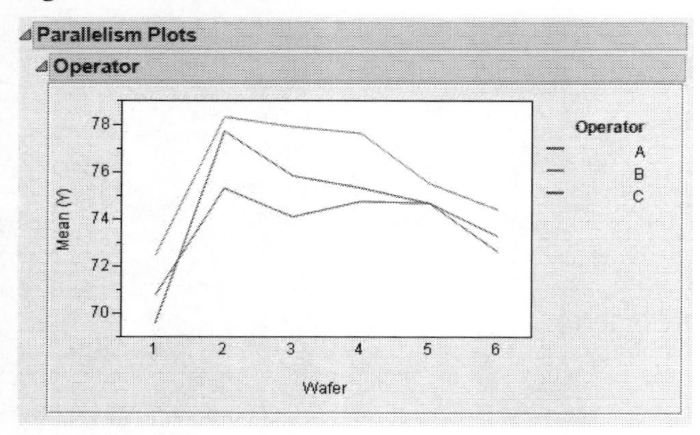

Looking at the parallelism plot by operator, you can see that the lines are relatively parallel and that there is only some minor crossing.

Examine Operator Consistency

Take a closer look at the variance between operators. From the red triangle menu next to Measurement Systems Analysis for Y, select **Test-Retest Error Comparison.**

Figure 7.10 Test-Retest Error Comparison

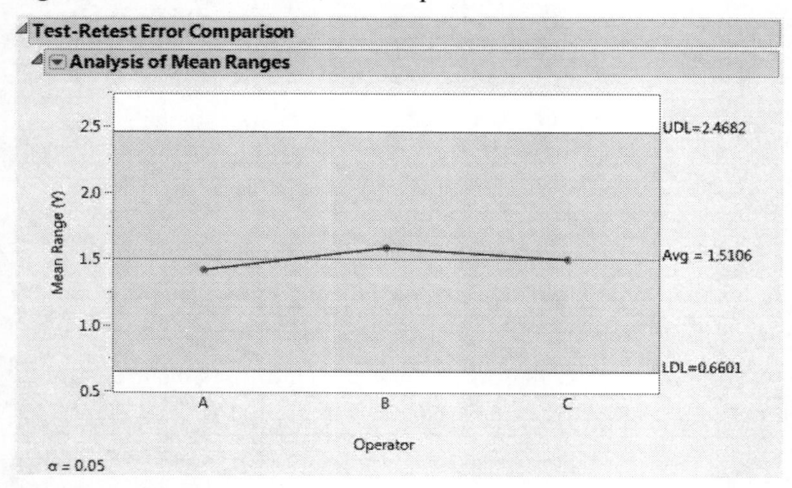

Looking at the Test-Retest Error Comparison, you can see that none of the operators have a test-retest error that is significantly different from the overall test-retest error. The operators appear to be measuring consistently.

Just to be sure, you decide to look at the Bias Comparison chart, which indicates whether an operator is measuring parts too high or too low. From the red triangle menu next to Measurement Systems Analysis for Y, select **Bias Comparison**.

Figure 7.11 Bias Comparison

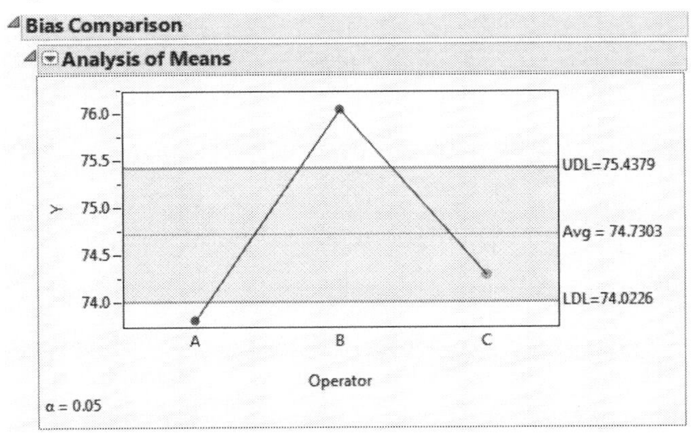

Looking at the Bias Comparison chart, you make the following observations:

- Operator A and Operator B have detectable measurement bias, as they are significantly different from the overall average.
- Operator A is significantly biased low.
- Operator B is significantly biased high.
- Operator C is not significantly different from the overall average.

Classify Your Measurement System

Examine the EMP Results report to classify your measurement system and look for opportunities for improvement. From the red triangle menu next to Measurement Systems Analysis for Y, select **EMP Results**.

Figure 7.12 EMP Results

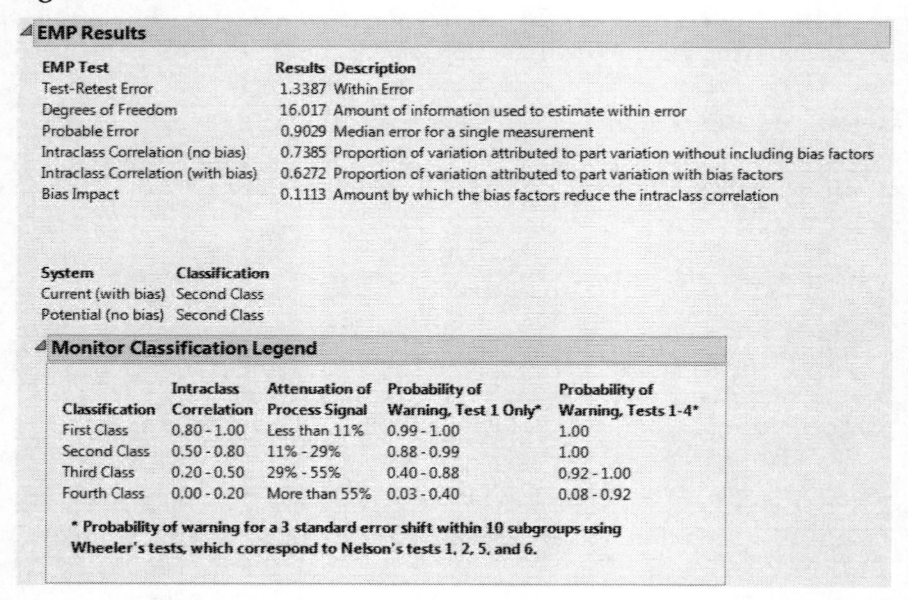

EMP Results

EMP Test	Results	Description
Test-Retest Error	1.3387	Within Error
Degrees of Freedom	16.017	Amount of information used to estimate within error
Probable Error	0.9029	Median error for a single measurement
Intraclass Correlation (no bias)	0.7385	Proportion of variation attributed to part variation without including bias factors
Intraclass Correlation (with bias)	0.6272	Proportion of variation attributed to part variation with bias factors
Bias Impact	0.1113	Amount by which the bias factors reduce the intraclass correlation

System	Classification
Current (with bias)	Second Class
Potential (no bias)	Second Class

Monitor Classification Legend

Classification	Intraclass Correlation	Attenuation of Process Signal	Probability of Warning, Test 1 Only*	Probability of Warning, Tests 1-4*
First Class	0.80 - 1.00	Less than 11%	0.99 - 1.00	1.00
Second Class	0.50 - 0.80	11% - 29%	0.88 - 0.99	1.00
Third Class	0.20 - 0.50	29% - 55%	0.40 - 0.88	0.92 - 1.00
Fourth Class	0.00 - 0.20	More than 55%	0.03 - 0.40	0.08 - 0.92

* Probability of warning for a 3 standard error shift within 10 subgroups using Wheeler's tests, which correspond to Nelson's tests 1, 2, 5, and 6.

The classification is Second Class, which means that there is a better than 88% chance of detecting a three standard error shift within ten subgroups, using Test one only. You notice that the bias factors have an 11% impact on the Intraclass Correlation. In other words, if you could eliminate the bias factors, your Intraclass Correlation coefficient would improve by 11%.

Explore the Ability of a Control Chart to Detect Process Changes

Use the Shift Detection Profiler to explore the probability that a control chart will be able to detect a change in your process. From the red triangle menu next to Measurement Systems Analysis for Y, select **Shift Detection Profiler**.

Figure 7.13 Shift Detection Profiler

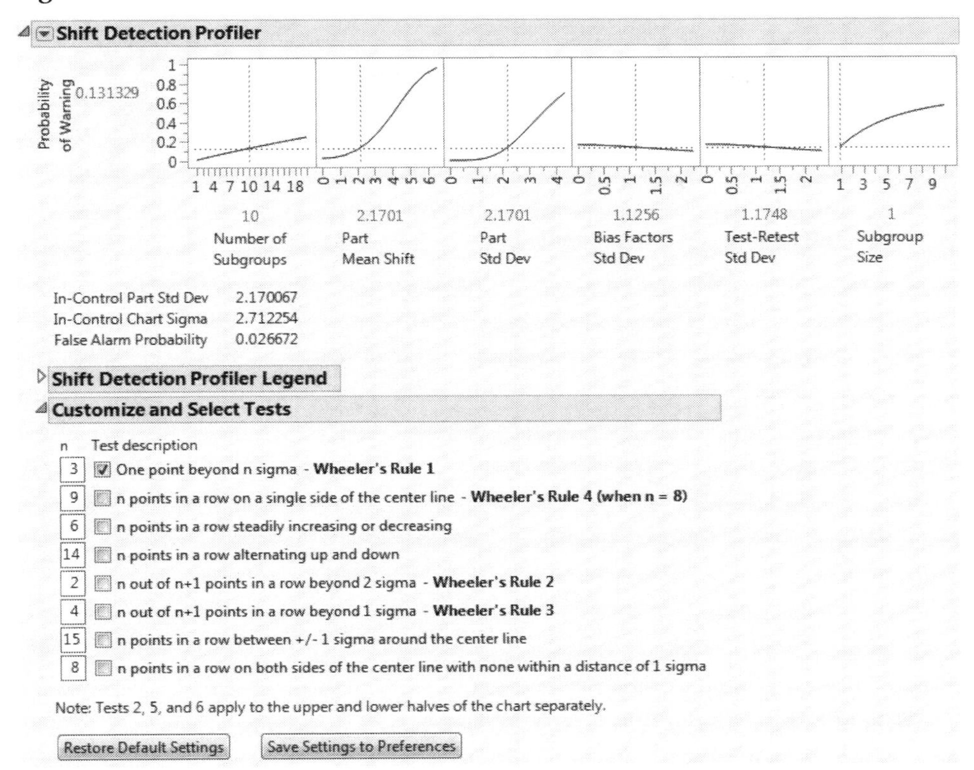

By default, the only test selected is for a point beyond the 3 sigma limits. Also note that the default Subgroup Size is 1, indicating that you are using an Individual Measurement chart.

Explore your ability to detect a shift in the mean of two part standard deviations in the 10 subgroups following the shift. Click the **Part Mean Shift** value of 2.1701 and change it to 4.34 (2.17 multiplied by 2). The probability of detecting a shift of twice the part standard deviation is 56.9%.

Next, see how eliminating bias affects your ability to detect the shift of two part standard deviations. Change the **Bias Factors Std Dev** value from 1.1256 to 0. The probability of detecting the shift increases to 67.8%.

Finally, add more tests to see how your ability to detect the two part standard deviation shift changes. In addition to the first test, select the second, fifth, and sixth tests (Wheeler's Rules 4, 2, and 3). With these four tests and no bias variation, your probability of detecting the shift is 99.9%.

You can also explore the effect of using a control chart based on larger subgroup sizes. For subgroup sizes of two or more, the control chart is an X-bar chart. Change the **Bias Factors Std**

Dev value back to 1.1256 and deselect all but the first test. Set the **Subgroup Size** in the profiler to 4. The probability of detecting the two part standard deviation shift is 98.5%.

Examine Measurement Increments

Finally, see how well your measurement increments are working. From the red triangle menu next to Measurement Systems Analysis for Y, select **Effective Resolution**.

Figure 7.14 Effective Resolution

Effective Resolution

Source		Value	Description
Probable Error	(PE)	0.9029	Median error for a single measurement
Current Measurement Increment	(MI)	0.01	Measurement increment estimated from data (in tenths)
Lower Bound Increment	(0.1*PE)	0.0903	Measurement increment should not be below this value
Smallest Effective Increment	(0.22*PE)	0.1986	Measurement increment is more effective above this value
Largest Effective Increment	(2.2*PE)	1.9865	Measurement increment is more effective below this value

Action: Drop a digit
Reason: The measurement increment of 0.01 is below the lowest measurement increment bound and should be adjusted to record fewer digits.

The Current Measurement Increment of 0.01 is below the Lower Bound Increment of 0.09, indicating that you should adjust your future measurements to record one less digit.

Statistical Details for Measurement Systems Analysis

Intraclass Correlation without bias is computed as follows:

$$r_{pe} = \frac{\hat{\sigma}_p^2}{\hat{\sigma}_p^2 + \hat{\sigma}_{pe}^2}$$

Intraclass Correlation with bias is computed as follows:

$$r_b = \frac{\hat{\sigma}_p^2}{\hat{\sigma}_p^2 + \hat{\sigma}_b^2 + \hat{\sigma}_{pe}^2}$$

Intraclass Correlation with bias and interaction factors is computed as follows:

$$r_{int} = \frac{\hat{\sigma}_p^2}{\hat{\sigma}_p^2 + \hat{\sigma}_b^2 + \hat{\sigma}_{int}^2 + \hat{\sigma}_{pe}^2}$$

Probable Error is computed as follows:

$$Z_{0.75} \times \hat{\sigma}_{pe}$$

Note the following:

$\hat{\sigma}_{pe}^2$ = variance estimate for pure error

$\hat{\sigma}_p^2$ = variance estimate for product

$\hat{\sigma}_b^2$ = variance estimate for bias factors

$\hat{\sigma}_{int}^2$ = variance estimate for interaction factors

$Z_{0.75}$ = the 75% quantile of standard normal distribution

Chapter 8

Variability Gauge Charts

Evaluate a Continuous Measurement Process Using Gauge R&R

Variability gauge charts analyze continuous measurements and can reveal how your measurement system is performing. You can also perform a gauge study to see measures of variation in your data.

Tip: This chapter covers only variability charts. For details about attribute charts, see the "Attribute Gauge Charts" chapter on page 199.

Figure 8.1 Example of a Variability Chart

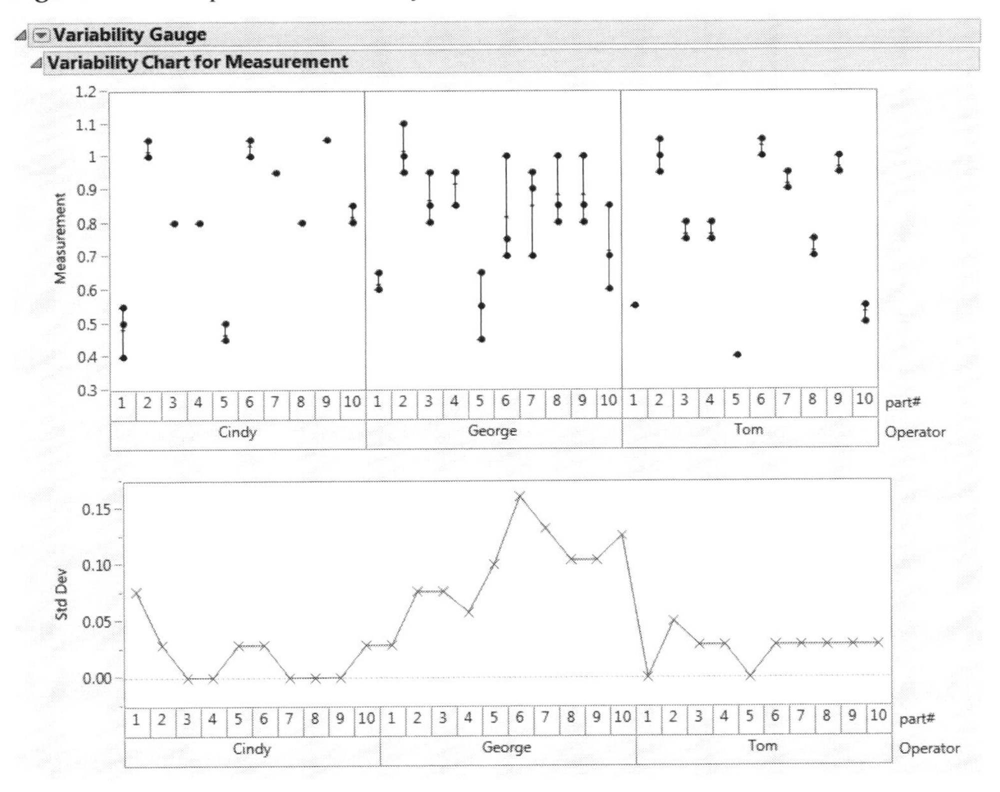

Overview of Variability Charts

Tip: The traditional name for a variability chart is a *multi vari* chart, but because that name is not well known, the more generic term variability chart is used instead.

Just as a control chart shows variation across time in a process, a variability chart shows the same type of variation across categories such as parts, operators, repetitions, and instruments. A variability chart plots the data and means for each level of grouping factors, with all plots side by side. Along with the data, you can view the mean, range, and standard deviation of the data in each category, seeing how they change across the categories. The report options are based on the assumption that the primary interest is how the mean and variance change across the categories.

Variability charts are commonly used for measurement systems analysis such as Gauge R&R. This analysis examines how much of the variability is due to operator variation (reproducibility) and measurement variation (repeatability). Gauge R&R is available for many combinations of crossed and nested models, regardless of whether the model is balanced.

Example of a Variability Chart

Suppose that you have data containing part measurements. Three operators, Cindy, George, and Tom, each took measurements of 10 parts. They measured each part three times, making a total of 90 observations. You want to identify the variation between operators.

1. Select **Help > Sample Data Library** and open Variability Data/2 Factors Crossed.jmp.
2. Select **Analyze > Quality and Process > Variability / Attribute Gauge Chart.**
3. For **Chart Type**, select **Variability**.
4. Select Measurement and click **Y, Response**.
5. Select Operator and click **X, Grouping**.
6. Select part# and click **Part, Sample ID**.
7. Click **OK**.

Figure 8.2 Example of a Variability Chart

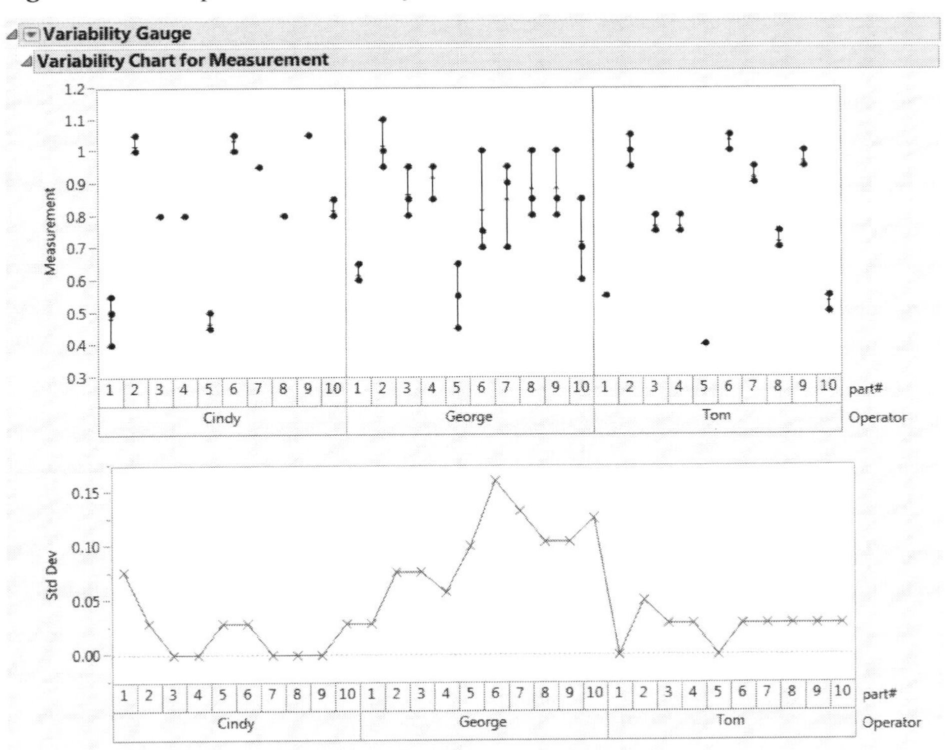

Looking at the Std Dev chart, you can see that Cindy and George have more variation in their measurements than Tom, who appears to be measuring parts the most consistently. George seems to have the most variation in his measurements, so he might be measuring parts the most inconsistently.

Launch the Variability/Attribute Gauge Chart Platform

Launch the Variability/Attribute Gauge Chart platform by selecting **Analyze > Quality and Process > Variability/Attribute Gauge Chart**. Set the **Chart Type** to **Variability**.

Figure 8.3 The Variability/Attribute Gauge Chart Launch Window

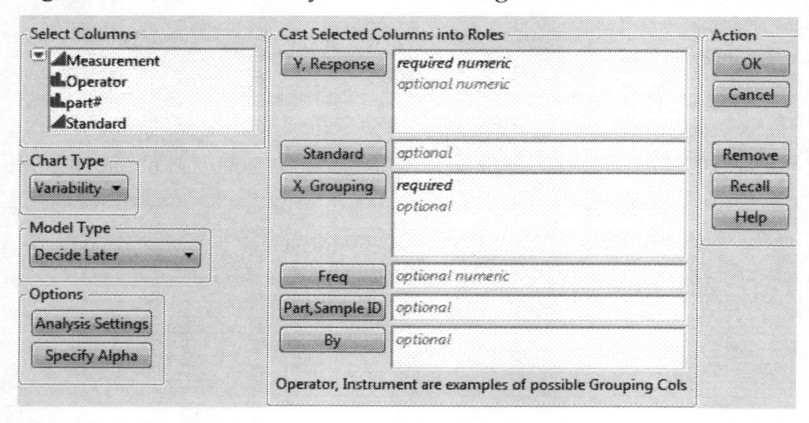

Chart Type Choose between a variability gauge analysis (for a continuous response) or an attribute gauge analysis (for a categorical response, usually "pass" or "fail").

Note: The content in this chapter covers only the **Variability** chart type. For details about the **Attribute** chart type, see the "Attribute Gauge Charts" chapter on page 199.

Model Type Choose the model type (**Main Effect**, **Crossed**, **Nested**, and so on). See "Statistical Details for Variance Components" on page 196.

Analysis Settings Specify the method for computing variance components. See "Analysis Settings" on page 184.

Specify Alpha Specify the alpha level used by the platform.

Y, Response Specify the measurement column. Specifying more than one *Y* column produces a separate variability chart for each response.

Standard Specify a standard or reference column that contains the "true" or known values for the measured part. Including this column enables the **Bias** and **Linearity Study** options. These options perform analysis on the differences between the observed measurement and the reference or standard value. See "Bias Report" on page 190 and "Linearity Study" on page 190.

X, Grouping Specify the classification columns that group the measurements. If the factors form a nested hierarchy, specify the higher terms first. If you are doing a gauge study, specify the operator first and then the part.

Freq Identifies the data table column whose values assign a frequency to each row. Can be useful when you have summarized data.

Part, Sample ID Identifies the part or sample that is being measured.

By Identifies a column that creates a report consisting of separate analyses for each level of the variable.

For more information about the launch window, see the Get Started chapter in the *Using JMP* book.

The Variability Gauge Chart

The variability chart and the standard deviation chart show patterns of variation. You can use these charts to identify possible groups of variation (within subgroups, between subgroups, over time). If you notice that any of these sources of variation are large, you can then work to reduce the variation for that source.

Figure 8.4 Variability Gauge Chart

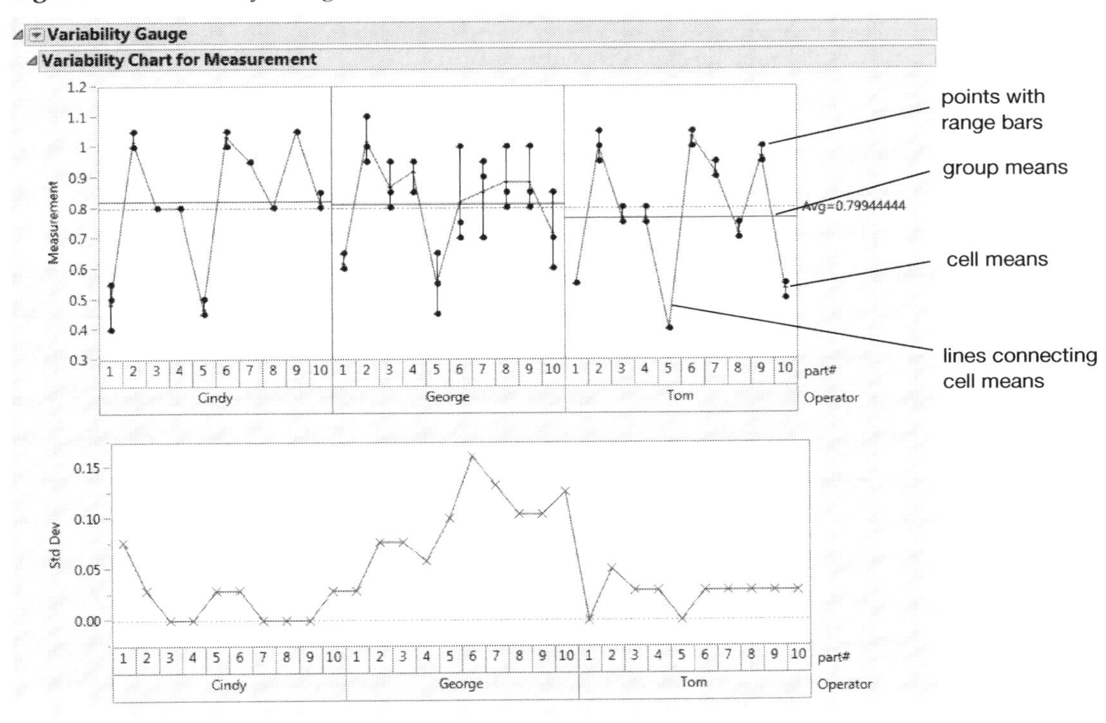

The charts show the response on the *y*-axis and a multilevel, categorized *x*-axis.

In Figure 8.4, the Measurement chart shows the range of measurements for each operator by part. Each measurement appears on the chart. Maximum and minimum bars indicate the range of values for each cell, and a cell means bar indicates the median value for each combination of values. The Std Dev chart plots the standard deviation of the measurements taken on each part by operator.

You can add features to the charts, as illustrated in Figure 8.4. See "Variability Gauge Platform Options" on page 180.

To replace variables in charts, do one of the following:

- Swap existing variables by dragging a variable from one axis label to the other axis label. When you drag a variable over a chart or click on an axis label, the axis labels are highlighted. This indicates where to drop the variable.
- Click on a variable in the Columns panel of the associated data table and drag it onto an axis label.

In other platforms, rows that are excluded in the associated data table still appear on the charts or plots. However, in variability charts, excluded rows do not appear on the charts.

Variability Gauge Platform Options

Use the red triangle options to modify the appearance of the chart, perform Gauge R&R analysis, and compute variance components.

Note: Figure 8.4 illustrates some of these options.

Tip: To set the default behavior of these options, select **File > Preferences > Platforms > Variability Chart**.

Vertical Charts Changes the layout to horizontal or vertical.

Variability Chart Shows or hides the variability chart.

Show Points Shows or hides the points for individual rows.

Show Range Bars Shows or hides the bars indicating the minimum and the maximum value of each cell.

Show Cell Means Shows or hides the mean mark for each cell.

Connect Cell Means Connects or disconnects cell means within a group of cells.

Show Separators Shows or hides the separator lines between levels of the **X, Grouping** variables.

Show Group Means (Available only if you have two or more X, Grouping variables or one X, Grouping variable and one Part, Sample ID variable) Shows or hides the mean for groups of cells, represented by a horizontal solid line. A window appears, prompting you to select one of the grouping variables.

Show Grand Mean Shows or hides the overall mean, represented by a gray dotted line across the entire graph.

Show Grand Median Shows or hides the overall median, represented by a blue dotted line across the entire graph.

Show Box Plots Shows or hides box plots.

Mean Diamonds Shows or hides mean diamonds. The confidence intervals use the within-group standard deviation for each cell.

XBar Control Limits Shows or hides lines at the UCL and LCL on the variability chart.

Points Jittered Adds some random noise to the plotted points so that coincident points do not plot on top of one another.

Show Bias Line (Available only if you have specified a **Standard** variable) Shows or hides the bias line in the main variability chart.

Show Standard Mean (Available only if you have specified a **Standard** variable) Shows or hides the mean of the standard column.

Variability Summary Report Shows or hides a report that gives the mean, standard deviation, standard error of the mean, lower and upper confidence intervals, and the minimum, maximum, and number of observations.

Std Dev Chart Shows or hides a separate graph that shows cell standard deviations across category cells.

Mean of Std Dev Shows or hides a line at the mean standard deviation on the Std Dev chart.

S Control Limits Shows or hides lines showing the LCL and UCL in the Std Dev chart.

Group Means of Std Dev Shows or hides the mean lines on the Std Dev chart.

Heterogeneity of Variance Tests Performs a test for comparing variances across groups. See "Heterogeneity of Variance Tests" on page 182.

Variance Components Estimates the variance components for a specific model. Variance components are computed for these models: main effects, crossed, nested, crossed then nested (three factors only), and nested then crossed (three factors only). See "Variance Components" on page 183.

Gauge Studies Contains the following options:

- **Gauge RR** interprets the first factors as grouping columns and the last factor as Part, and creates a gauge R&R report using the estimated variance components. (Note that there is also a Part field in the launch window). See "Gauge RR Option" on page 186.

- **Discrimination Ratio** characterizes the relative usefulness of a given measurement for a specific product. It compares the total variance of the measurement with the variance of the measurement error. See "Discrimination Ratio" on page 189.
- **Misclassification Probabilities** show probabilities for rejecting good parts and accepting bad parts. See "Misclassification Probabilities" on page 189.
- **Bias Report** shows the average difference between the observed values and the standard. A graph of the average biases and a summary table appears. This option is available only when you specify a Standard variable in the launch window. See "Bias Report" on page 190.
- **Linearity Study** performs a regression using the standard values as the X variable and the bias as the Y variable. This analysis examines the relationship between bias and the size of the part. Ideally, you want the slope to equal 0. A nonzero slope indicates that your gauge performs differently with different sized parts. This option is available only when you specify a Standard variable in the launch window. See "Linearity Study" on page 190.
- **Gauge RR Plots** shows or hides **Mean Plots** (the mean response by each main effect in the model) and **Std Dev** plots. If the model is purely nested, the graphs appear with a nesting structure. If the model is purely crossed, interaction graphs appear. Otherwise, the graphs plot independently at each effect. For the standard deviation plots, the red lines connect $\sqrt{\text{mean weighted variance}}$ for each effect.
- **AIAG Labels** enables you to specify that quality statistics should be labeled in accordance with the AIAG standard, which is used extensively in automotive analyses.

See the JMP Reports chapter in the *Using JMP* book for more information about the following options:

Local Data Filter Shows or hides the local data filter that enables you to filter the data used in a specific report.

Redo Contains options that enable you to repeat or relaunch the analysis. In platforms that support the feature, the Automatic Recalc option immediately reflects the changes that you make to the data table in the corresponding report window.

Save Script Contains options that enable you to save a script that reproduces the report to several destinations.

Save By-Group Script Contains options that enable you to save a script that reproduces the platform report for all levels of a By variable to several destinations. Available only when a By variable is specified in the launch window.

Heterogeneity of Variance Tests

Note: See "Example of the Heterogeneity of Variance Test" on page 191.

The **Heterogeneity of Variance Tests** option performs a test for comparing variances across groups. The test is an Analysis of Means for Variances (ANOMV) based method. This method indicates whether any of the group standard deviations are different from the square root of the average group variance.

To be robust against non-normal data, the method uses a permutation simulation to compute decision limits. For complete details about this method, see Wludyka and Sa (2004). Because the method uses simulations, the decision limits can be slightly different each time. To obtain the same results each time, hold down CTRL and SHIFT and select the option, and then specify the same random seed.

The red triangle menus for the test reports contain the following options:

Set Alpha Level Sets the alpha level for the test.

Show Summary Report Shows or hides a summary report for the test. The values in the report are the same values that are shown in the plot.

Note: The values in the plots and the Summary Reports are the values used in performing the test, not the group standard deviations.

Display Options Shows or hides the decision limits, shading, center line, and needles.

Variance Components

The **Variance Components** option models the variation from measurement to measurement. The response is assumed to be a constant mean plus random effects associated with various levels of the classification.

Note: Once you select the **Variance Components** option, if you did not select the **Model Type** in the launch window (if you selected **Decide Later**), you are prompted to select the model type. For more information about model types, see "Launch the Variability/Attribute Gauge Chart Platform" on page 178.

Figure 8.5 Example of the Variance Components Report

Analysis of Variance

Source	DF	SS	Mean Square	F Ratio	Prob > F
Operator	2	0.054889	0.02744	1.3150	0.2931
part#	9	2.633583	0.29262	14.0209	<.0001*
Operator*part#	18	0.375667	0.02087	5.0425	<.0001*
Within	60	0.248333	0.00414		
Total	89	3.312472	0.03722		

Variance Components

Component	Var Component	% of Total	20 40 60 80	Sqrt(Var Comp)
Operator	0.00021914	0.5461		0.01480
part#	0.03019444	75.2		0.17377
Operator*part#	0.00557716	13.9		0.07468
Within	0.00413889	10.3		0.06433
Total	0.04012963	100.0		0.20032

The Analysis of Variance report appears only if the EMS method of variance component estimation is used. This report shows the significance of each effect in the model.

The Variance Components report shows the estimates themselves. See "Statistical Details for Variance Components" on page 196.

Analysis Settings

From the launch window, click **Analysis Settings** to choose the method for computing variance components.

Figure 8.6 Analysis Settings Window

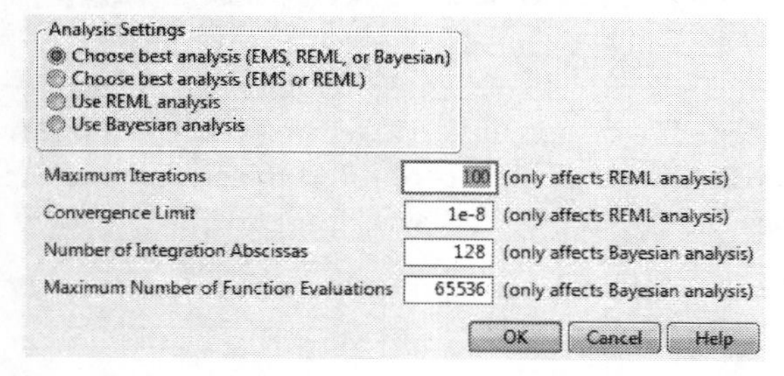

Choose best analysis (EMS, REML, or Bayesian) Chooses the best analysis from EMS, REML, or Bayesian, using the following logic:

- If the data are balanced, and if no variance components are negative, the EMS (expected mean squares) method is used to estimate the variance components.
- If the data are unbalanced, the REML (restricted maximum likelihood) method is used, unless a variance component is estimated to be negative, then the Bayesian method is used.
- If any variance component is estimated to be negative using the EMS method, the Bayesian method is used.

- If there is confounding in the variance components, then the bounded REML method is used, and any negative variance component estimates are set to zero.

Choose best analysis (EMS or REML) Chooses the best analysis from EMS or REML, using the same logic as the **Choose best analysis (EMS, REML, or Bayesian)** option. However, this option never uses the Bayesian method, even for negative variance components. The bounded REML method is used and any negative variance component is forced to be 0.

Use REML analysis Uses the bounded REML method, even if the data are balanced. The bounded REML method can handle unbalanced data and forces any negative variance component to be 0.

Use Bayesian analysis Uses the Bayesian method. The Bayesian method can handle unbalanced data and forces all variances components to be positive and nonzero. If there is confounding in the variance components, then the bounded REML method is used, and any negative variance component estimates are set to zero. The method implemented in JMP computes the posterior means using a modified version of Jeffreys' prior. For details, see Portnoy (1971) and Sahai (1974).

Maximum Iterations (Applicable only for the REML method) For difficult problems, you might want to increase the number of iterations. Increasing this value means that JMP will try more times to find a solution in the optimization phase.

Convergence Limit (Applicable only for the REML method) For problems where you want greater precision, you might want to change the convergence limit to be smaller. Decreasing this value means that JMP will find the solution to a higher level of accuracy in the optimization phase. However, this can increase the time taken to find a solution. Providing a larger convergence value returns quicker results, but is less precise.

Number of Iteration Abscissas (Applicable only for the Bayesian method) For greater accuracy, you might want to increase the number of iteration abscissas. However, this can increase the time taken to find a solution. Providing a smaller number returns quicker results, but is less precise.

Maximum Number of Function Evaluations (Applicable only for the Bayesian method) For greater accuracy, you might want to increase the maximum number of function evaluations. However, this can increase the time taken to find a solution. Providing a smaller number returns quicker results, but is less precise.

About the Gauge R&R Method

The Gauge R&R method analyzes how much of the variability in your measurement system is due to operator variation (reproducibility) and measurement variation (repeatability). Gauge R&R studies are available for many combinations of crossed and nested models, regardless of whether the model is balanced.

Tip: Alternatively, you can use the EMP method to assess your measurement system. See the "Measurement Systems Analysis" chapter on page 149.

Before performing a Gauge R&R study, you collect a random sample of parts over the entire range of part sizes from your process. Select several operators at random to measure each part several times. The variation is then attributed to the following sources:

- The *process variation*, from one part to another. This is the ultimate variation that you want to be studying if your measurements are reliable.
- The variability inherent in making multiple measurements, that is, *repeatability*. In Table 8.1 on page 186, this is called the *within variation*.
- The variability due to having different operators measure parts—that is, *reproducibility*.

A Gauge R&R analysis then reports the variation in terms of repeatability and reproducibility.

Table 8.1 Definition of Terms and Sums in Gauge R&R Analysis

Variances Sums	Term and Abbreviation	Alternate Term
V(Within)	Repeatability (EV)	Equipment Variation
V(Operator)+V(Operator*Part)	Reproducibility (AV)	Appraiser Variation
V(Operator*Part)	Interaction (IV)	Interaction Variation
V(Within)+V(Operator)+V(Operator*Part)	Gauge R&R (RR)	Measurement Variation
V(Part)	Part Variation (PV)	Part Variation
V(Within)+V(Operator)+V(Operator*Part)+V(Part)	Total Variation (TV)	Total Variation

A Shewhart control chart can identify processes that are going out of control over time. A variability chart can also help identify operators, instruments, or part sources that are systematically different in mean or variance.

Gauge RR Option

The **Gauge RR** option shows measures of variation interpreted for a gauge study of operators and parts.

Once you select the **Gauge RR** option, if you have not already selected the model type, you are prompted to do so. Then, modify the Gauge R&R specifications.

Note: The Platform preferences for Variability include the Gauge R&R Specification Dialog option. The preference is selected by default. Deselect the preference to use the spec limits that are defined in the data table.

Enter/Verify Gauge R&R Specifications

The Enter/Verify Gauge R&R Specifications window contains these options:

Choose tolerance entry method Choose how to enter the tolerance, as follows:

Select **Tolerance Interval** to enter the tolerance directly, where tolerance = USL – LSL.

Select **LSL and/or USL** to enter the specification limits and then have JMP calculate the tolerance.

K, Sigma Multiplier **K** is a constant value that you choose to multiply with sigma. For example, you might type 6 so that you are looking at 6*sigma or a 6 sigma process.

Tip: Modify the default value of **K** by selecting **File > Preferences > Platforms > Variability Chart**.

Tolerance Interval, USL-LSL Enter the tolerance for the process, which is the difference between the upper specification limits and the lower specification limits.

Spec Limits Enter upper and lower specification limits.

Historical Mean Computes the tolerance range for one-sided specification limits, either USL-Historical Mean or Historical Mean-LSL. If you do not enter a historical mean, the grand mean is used.

Historical Sigma Enter a value that describes the variation (you might have this value from history or past experience).

The Gauge R&R Report

Figure 8.7 Example of the Gauge R&R Report

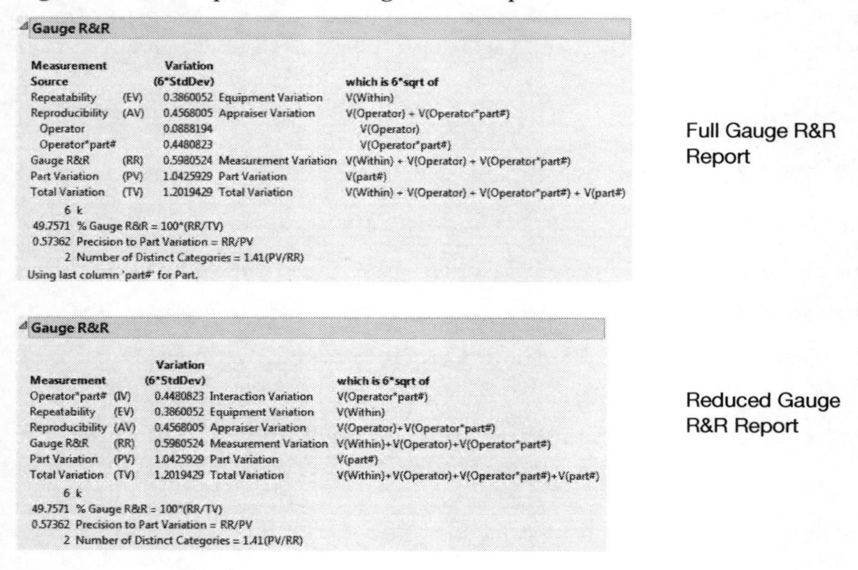

Gauge R&R

Measurement Source		Variation (6*StdDev)		which is 6*sqrt of
Repeatability	(EV)	0.3860052	Equipment Variation	V(Within)
Reproducibility	(AV)	0.4568005	Appraiser Variation	V(Operator) + V(Operator*part#)
Operator		0.0888194		V(Operator)
Operator*part#		0.4480823		V(Operator*part#)
Gauge R&R	(RR)	0.5980524	Measurement Variation	V(Within) + V(Operator) + V(Operator*part#)
Part Variation	(PV)	1.0425929	Part Variation	V(part#)
Total Variation	(TV)	1.2019429	Total Variation	V(Within) + V(Operator) + V(Operator*part#) + V(part#)

6 k
49.7571 % Gauge R&R = 100*(RR/TV)
0.57362 Precision to Part Variation = RR/PV
2 Number of Distinct Categories = 1.41(PV/RR)
Using last column 'part#' for Part.

Gauge R&R

Measurement		Variation (6*StdDev)		which is 6*sqrt of
Operator*part#	(IV)	0.4480823	Interaction Variation	V(Operator*part#)
Repeatability	(EV)	0.3860052	Equipment Variation	V(Within)
Reproducibility	(AV)	0.4568005	Appraiser Variation	V(Operator)+V(Operator*part#)
Gauge R&R	(RR)	0.5980524	Measurement Variation	V(Within)+V(Operator)+V(Operator*part#)
Part Variation	(PV)	1.0425929	Part Variation	V(part#)
Total Variation	(TV)	1.2019429	Total Variation	V(Within)+V(Operator)+V(Operator*part#)+V(part#)

6 k
49.7571 % Gauge R&R = 100*(RR/TV)
0.57362 Precision to Part Variation = RR/PV
2 Number of Distinct Categories = 1.41(PV/RR)

Note: To generate the reduced Gauge R&R report, select **File > Preferences > Platforms > Variability Chart** > **Reduced Gauge RR Report.**

In this example, the values in the Variation column are the square roots of sums of variance components scaled by the value of *k* (6 in this example).

Table 8.2 shows guidelines for measurement variation, as suggested by Barrentine (1991).

Table 8.2 Acceptable Percent Measurement Variation

< 10%	excellent
11% to 20%	adequate
21% to 30%	marginally acceptable
> 30%	unacceptable

Note the following:

- If you have provided a **Tolerance Interval** in the Enter/Verify Gauge R&R Specifications window, a % of Tolerance column appears in the Gauge R&R report. This column is computed as 100*(Variation/Tolerance). Also, a Precision-to-Tolerance ratio appears at the bottom of the report. This ratio represents the proportion of the tolerance or capability interval that is lost due to gauge variability.

- If you have provided a **Historical Sigma** in the Enter/Verify Gauge R&R Specifications window, a % Process column appears in the Gauge R&R report. This column is defined as follows: 100*(Variation/(K*Historical Sigma)).
- The Number of Distinct Categories (NDC) is defined as (1.41*(PV/RR)), rounded down to the nearest integer.

Discrimination Ratio

The discrimination ratio characterizes the relative usefulness of a given measurement for a specific product. Generally, when the discrimination ratio is less than 2, the measurement cannot detect product variation, implying that the measurement process needs improvement. A discrimination ratio greater than 4 adequately detects unacceptable product variation, implying that the production process needs improvement.

See "Statistical Details for the Discrimination Ratio" on page 197 for more information.

Misclassification Probabilities

Due to measurement variation, good parts can be rejected and bad parts can be accepted. This is called misclassification. Once you select the **Misclassification Probabilities** option, if you have not already done so, you are prompted to select the model type and enter specification limits.

Figure 8.8 Example of the Misclassification Probabilities Report

Misclassification Probabilities

Description	Probability
P(Good part is falsely rejected)	0.0802
P(Bad part is falsely accepted)	0.2787
P(Part is good and is rejected)	0.0735
P(Part is bad and is accepted)	0.0235
P(Part is good)	0.9157

Note the following:

- The first and second values are conditional probabilities.
- The third and fourth values are joint probabilities.
- The fifth value is a marginal probability.
- The first four values are probabilities of errors that decrease as the measurement variation decreases.

Bias Report

The **Bias Report** shows a graph for Overall Measurement Bias with a summary table and a graph for Measurement Bias by Standard with a summary table. The average bias, or the differences between the observed values and the standard values, appears for each level of the X variable. A *t* test for the bias is also given.

The **Bias Report** option is available only when a Standard variable is provided in the launch window.

The Measurement Bias Report contains the following red triangle options:

Confidence Intervals Calculates confidence intervals for the average bias for each part and places marks on the Measurement Bias Report by Standard plot.

Measurement Error Graphs Produces a graph of Measurement Error versus all grouping columns together. There are also graphs of Measurement Error by each grouping column separately.

Linearity Study

The **Linearity Study** performs a regression analysis using the standard variable as the X variable and the bias as the Y variable. This analysis examines the relationship between bias and the size of the part. Ideally, you want to find a slope of zero. If the slope is significantly different from zero, you can conclude that there is a significant relationship between the size of the part or variable measured as a standard and the ability to measure.

The **Linearity Study** option is available only when a Standard variable is provided in the launch window.

The report shows the following information:

- Bias summary statistics for each standard.
- An ANOVA table that tests if the slope of the line is equal to zero.
- The line parameters, including tests for the slope (linearity) and intercept (bias). The test for the intercept is useful only if the test on the slope fails to reject the hypothesis of slope = 0.
- The equation of the line appears directly beneath the graph.

The Linearity Study report contains the following red triangle options:

Set Alpha Level Changes the alpha level that is used in the bias confidence intervals.

Linearity by Groups Produces separate linearity plots for each level of the **X, Grouping** variables that you specified in the launch window.

Additional Examples of Variability Charts

This section contains additional examples of variability charts.

Example of the Heterogeneity of Variance Test

Suppose that you have data containing part measurements. Three operators (Cindy, George, and Tom) each took measurements of 10 parts. They measured each part three times, making a total of 90 observations. You want to examine the following:

- whether the variance of measurements for each operator are the same or different
- whether the variance for each part is the same or different
- whether the variance for each Operator*part combination is the same or different

Ideally, you want all of the variances for each of the groups to be considered the same statistically.

1. Select **Help > Sample Data Library** and open Variability Data/2 Factors Crossed.jmp.
2. Select **Analyze > Quality and Process > Variability / Attribute Gauge Chart.**
3. Select Measurement and click **Y, Response.**
4. Select Operator and click **X, Grouping.**
5. Select part# and click **Part, Sample ID.**
6. Set the **Chart Type** to **Variability.**
7. Click **OK.**
8. From the red triangle menu, select **Heterogeneity of Variance Tests.**
9. Select **Crossed.**
10. Click **OK.**

    You see a message that JMP will change your ordinal effect to a nominal one. Click **OK** to dismiss it.

Figure 8.9 Heterogeneity of Variances Tests Report

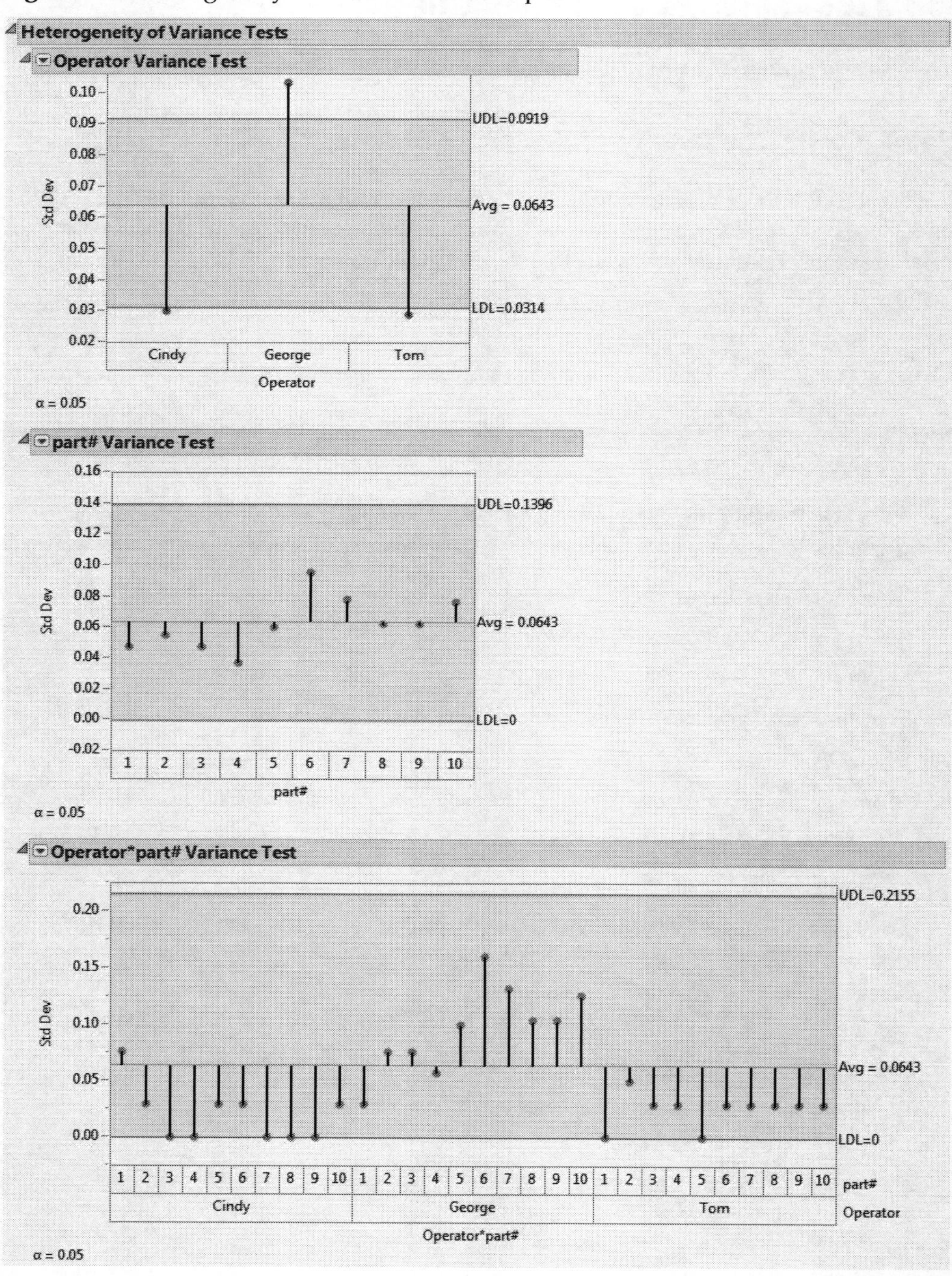

Note: Because the method uses simulations, the decision limits can be slightly different each time.

In the Operator Variance test, all three levels exceed the upper and lower decision limits. From this, you conclude that each operator has a different variability from the square root of the average group variance. You might want to examine why the variation between each operator is different.

For the part# Variance test and the interaction (Operator*part#) Variance test, none of the levels exceed the decision limits. From this, you conclude that the variances are not statistically different from the square root of the average group variance. Each part has a similar variance to the other parts, and each Operator*part# combination has similar variance to the other Operator*part# combinations.

Example of the Bias Report Option

Note: This data comes from the Automotive Industry Action Group (AIAG) (2002), *Measurement Systems Analysis Reference Manual*, 3rd edition, 94.

Assume that as a plant supervisor, you are introducing a new measurement system into your process. As part of the Production Part Approval Process (PPAP), the bias and linearity of the measurement system needs to be evaluated. Five parts were chosen throughout the operating range of the measurement system, based on documented process variation. Each part was measured by layout inspection to determine its reference value. Each part was then measured twelve times by the lead operator. The parts were selected at random during the day. In this example, you want to examine the overall bias and the individual measurement bias (by standard).

1. Select **Help > Sample Data Library** and open Variability Data/MSALinearity.jmp.
2. Select **Analyze > Quality and Process > Variability / Attribute Gauge Chart.**
3. Select Response and click **Y, Response.**
4. Select Standard and click **Standard.**
5. Select Part and click **X, Grouping.**
6. Set the **Chart Type** to **Variability.**
7. Click **OK.**
8. From the red triangle menu, select **Gauge Studies > Bias Report.**

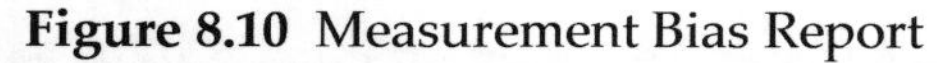

Figure 8.10 Measurement Bias Report

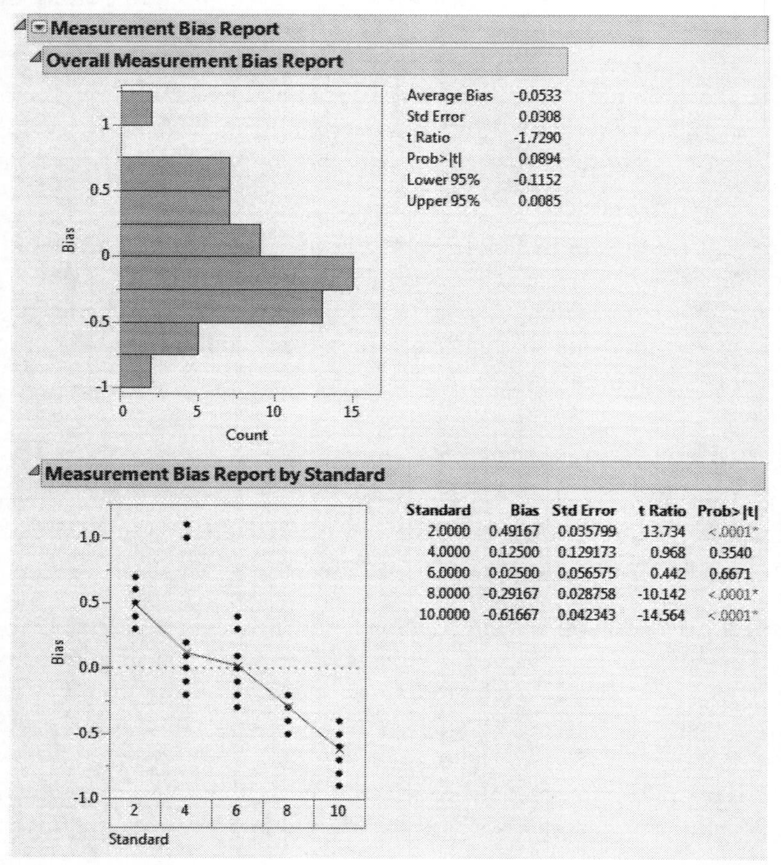

Average Bias	-0.0533
Std Error	0.0308
t Ratio	-1.7290
Prob>|t|	0.0894
Lower 95%	-0.1152
Upper 95%	0.0085

Standard	Bias	Std Error	t Ratio	Prob>|t|
2.0000	0.49167	0.035799	13.734	<.0001*
4.0000	0.12500	0.129173	0.968	0.3540
6.0000	0.02500	0.056575	0.442	0.6671
8.0000	-0.29167	0.028758	-10.142	<.0001*
10.0000	-0.61667	0.042343	-14.564	<.0001*

The bias (**Response** minus **Standard**) is calculated for every measurement. The Overall Measurement Bias Report shows a histogram of the bias and a t-test to see whether the average bias is equal to 0. You can see that the Average Bias is not zero, it is -0.0533. However, zero is contained within the confidence interval (-0.1152,0.0085), which means that the Average Bias is not significantly different from 0. Using a significance level of 0.05, you can see that the *p*-value is greater than 0.05, which also shows that the Average Bias is not significantly different from 0.

The Measurement Bias Report by Standard shows average bias values for each part. The bias averages are plotted on the graph along with the actual bias values for every part, so that you can see the spread. In this example, part number 1 (with a standard value of 2) is biased high and parts 4 and 5 (with standard values of 8 and 10) are biased low.

Tip: To see confidence intervals for the bias, right-click in the table and select **Columns > Lower 95% and Upper 95%**.

Example of a Linearity Study

Using the same data and scenario as the **Bias Report** option, you can now examine the linearity to determine whether there is a significant relationship between the size of the parts and the operator's ability to measure them.

1. Select **Help > Sample Data Library** and open Variability Data/MSALinearity.jmp.
2. Select **Analyze > Quality and Process > Variability / Attribute Gauge Chart.**
3. Select Response and click **Y, Response.**
4. Select Standard and click **Standard.**
5. Select Part and click **X, Grouping.**
6. Set the **Chart Type** to **Variability.**
7. Click **OK.**
8. From the red triangle menu, select **Gauge Studies > Linearity Study.**
9. In the window that prompts you to Specify Process Variation, type 16.5368.

Figure 8.11 Linearity Study

Linearity Study

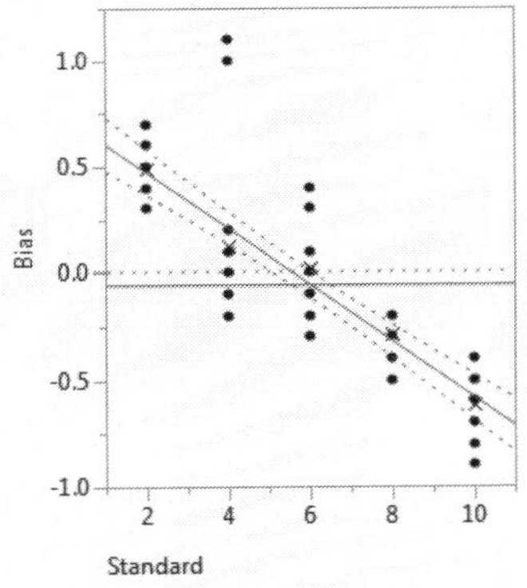

Bias = 0.7366667 - 0.1316667 Standard

Standard	Avg Response	Bias	Lower CL	Upper CL
2.0000	2.49167	0.49167	0.36612	0.58055
4.0000	4.12500	0.12500	0.13419	0.28581
6.0000	6.02500	0.02500	-0.11524	0.00857
8.0000	7.70833	-0.29167	-0.39248	-0.24085
10.0000	9.38333	-0.61667	-0.68722	-0.47278

Source	DF	Sum of Squares	Mean Square	F Ratio
Model	1	8.321333	8.32133	145.0232
Error	58	3.328000	0.05738	Prob > F
C. Total	59	11.649333		<.0001*

	Estimate	Std Error	t Ratio	Prob>|t|
Intercept	0.73667	0.072524	10.158	<.0001*
Slope	-0.13167	0.010933	-12.043	<.0001*

		Which equals
Linearity	-2.177	Slope * Process Variation
% Linearity	13.167	100 * abs(Slope)
Avg Bias	-0.053	Bias averaged over all parts
% Bias	0.323	100 * abs(Avg Bias) / Process Variation
Process Variation	16.537	Entered on dialog
R-Squared	0.714	
Alpha	0.050	

Note the following:

- The slope is -0.131667. This value appears as part of the equation below the graph, and also in the third table.

- The *p*-value associated with the test on the slope is quite small, <.0001. The *t* test for the slope is testing whether the bias changes with the standard value.

Because the *p*-value is small, you can conclude that there is a significant linear relationship between the size of the parts and the operator's ability to measure them. You can also see this in the graph. If the part or standard value is small, the bias is high, and vice versa.

Statistical Details for Variability Charts

This section contains statistical details for variance components and the discrimination ratio.

Statistical Details for Variance Components

The exact model type that you choose depends on how the data was collected. For example, are the operators measuring the same parts (in which case you have a crossed design) or are they measuring different parts (in which case you have a nested design)? To illustrate, in a model where *B* is nested within *A*, multiple measurements are nested within both *B* and *A*, and there are *na•nb•nw* measurements, as follows:

- *na* random effects are due to *A*
- *na•nb* random effects due to each *nb B* levels within *A*
- *na•nb•nw* random effects due to each *nw* levels within *B* within *A*:

$$y_{ijk} = u + Za_i + Zb_{ij} + Zw_{ijk}.$$

The *Z*s are the random effects for each level of the classification. Each *Z* is assumed to have a mean of zero and to be independent from all other random terms. The variance of the response *y* is the sum of the variances due to each *z* component:

$$\text{Var}(y_{ijk}) = \text{Var}(Za_i) + \text{Var}(Zb_{ij}) + \text{Var}(Zw_{ijk}).$$

Table 8.3 shows the supported models and what the effects in the model would be.

Table 8.3 Models Supported by the Variability Charts Platform

Model	Factors	Effects in the Model
Main Effects	1	A
	2	A, B
	unlimited	and so on, for more factors

Table 8.3 Models Supported by the Variability Charts Platform *(Continued)*

Model	Factors	Effects in the Model
Crossed	1	A
	2	A, B, A*B
	3	A, B, A*B, C, A*C, B*C, A*B*C
	4	A, B, A*B, C, A*C, B*C, A*B*C, D, A*D, B*D, A*B*D, C*D, A*C*D, B*C*D, A*B*C*D,
	unlimited	and so on, for more factors
Nested	1	A
	2	A, B(A)
	3	A, B(A), C(A,B)
	4	A, B(A), C(A,B), D(A,B,C)
	unlimited	and so on, for more factors
Crossed then Nested	3	A, B, A*B, C(A,B)
Nested then Crossed	3	A, B(A), C, A*C, C*B(A)

Statistical Details for the Discrimination Ratio

The discrimination ratio compares the total variance of the measurement, M, with the variance of the measurement error, E. The discrimination ratio is computed for all main effects, including nested main effects. The discrimination ratio, D, is computed as follows:

$$D = \sqrt{2\left(\frac{P}{T-P}\right)+1}$$

where:

P = estimated variance for a factor

T = estimated total variance

Chapter 9

Attribute Gauge Charts

Evaluate a Categorical Measurement Process Using Agreement Measures

Attribute charts analyze categorical measurements and can help show you measures of agreement across responses, such as raters. In *attribute data,* the variable of interest has a finite number of categories. Typically, data has only two possible results, such as pass or fail. You can examine aspects such as how effective raters were at classifying a part, how much they agreed with each other, and how much they agreed with themselves over the course of several ratings.

Tip: This chapter covers only attribute charts. For details about variability charts, see the "Variability Gauge Charts" chapter on page 175.

Figure 9.1 Example of an Attribute Chart

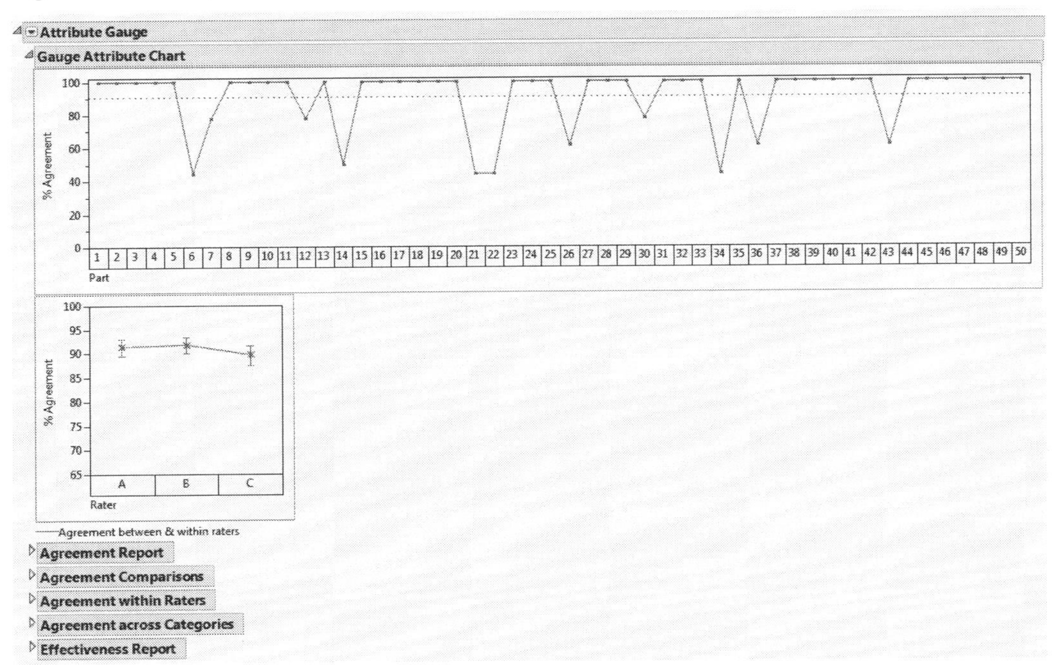

Attribute Gauge Charts Overview

Before you create an attribute gauge chart, your data should be formatted using the following guidelines:

- In order to compare agreement among raters, each rater in the data table must be in a separate column. These columns are then assigned to the **Y, Response** role in the launch window. In Figure 9.2, each rater (A, B, and C) is in a separate column.
- Responses in the different columns can be character (pass or fail) or numeric (0 or 1). In Figure 9.2, rater responses are numeric (0 for pass, 1 for fail). All response columns must have the same data type.
- Any other variables of interest that you might want to use as **X, Grouping** variables should appear stacked in one column each (see the Part column in Figure 9.2). You can also define a Standard column, which produces reports that compare raters with the standard. The Standard column and response columns must have the same data type.

Figure 9.2 Attribute Gauge Data

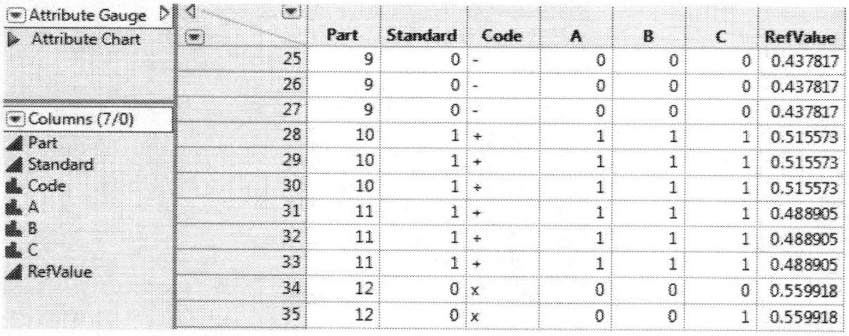

Attribute Gauge
Attribute Chart
Columns (7/0)
Part
Standard
Code
A
B
C
RefValue

	Part	Standard	Code	A	B	C	RefValue
25	9	0	-	0	0	0	0.437817
26	9	0	-	0	0	0	0.437817
27	9	0	-	0	0	0	0.437817
28	10	1	+	1	1	1	0.515573
29	10	1	+	1	1	1	0.515573
30	10	1	+	1	1	1	0.515573
31	11	1	+	1	1	1	0.488905
32	11	1	+	1	1	1	0.488905
33	11	1	+	1	1	1	0.488905
34	12	0	x	0	0	0	0.559918
35	12	0	x	0	0	1	0.559918

Example of an Attribute Gauge Chart

Suppose that you have data containing pass or fail ratings for parts. Three raters, identified as A, B, and C, each noted a 0 (pass) or a 1 (fail) for 50 parts, three times each. You want to examine how effective the raters were in correctly classifying the parts, and how well the raters agreed with each other and with themselves over the course of the ratings.

1. Select **Help > Sample Data Library** and open Attribute Gauge.jmp.
2. Select **Analyze > Quality and Process > Variability / Attribute Gauge Chart.**
3. For **Chart Type**, select **Attribute.**
4. Select A, B, and C and click **Y, Response.**
5. Select Standard and click **Standard.**
6. Select Part and click **X, Grouping.**

7. Click **OK**.

Figure 9.3 Example of an Attribute Chart

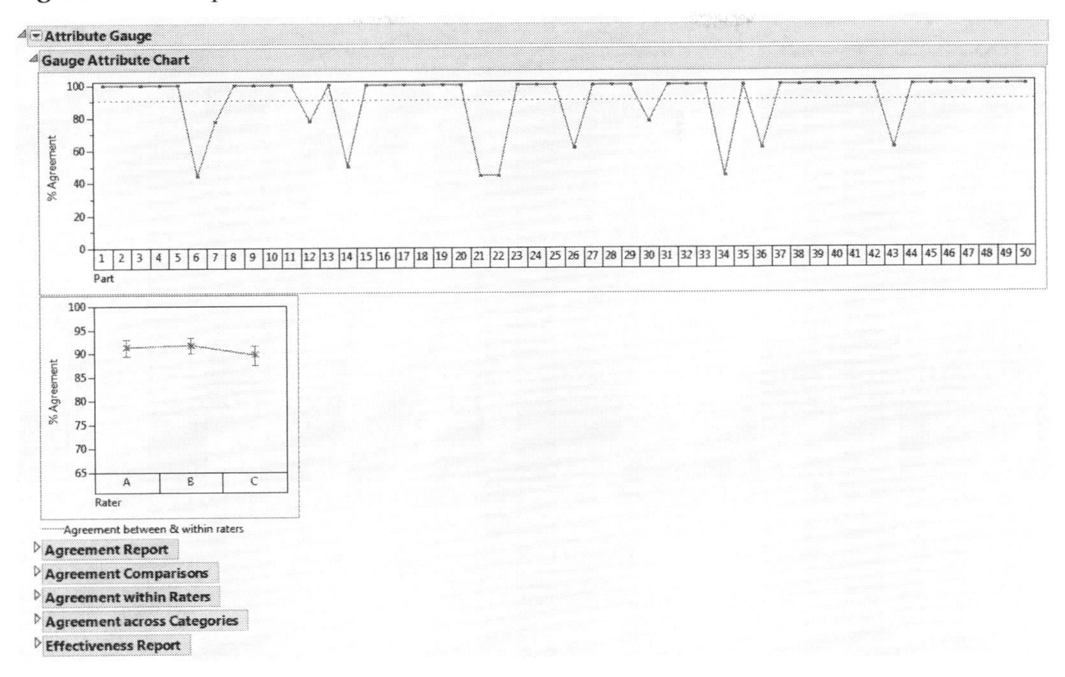

The first chart (Part) shows how well the raters agreed with each other for each part. For example, here you can see that the percent agreement dropped for part 6, 12, 14, 21, 22, and so on. These parts might have been more difficult to categorize.

The second chart (Rater) shows each rater's agreement with him or herself and the other raters for a given part, summed up over all of the parts. In this example, it looks like the performance of the raters is relatively similar. Rater C had the lowest agreement, but the difference is not major (about 89% instead of 91%).

Launch the Variability/Attribute Gauge Chart Platform

Launch the Variability/Attribute Gauge Chart platform by selecting **Analyze > Quality and Process > Variability/Attribute Gauge Chart**. Set the **Chart Type** to **Attribute**.

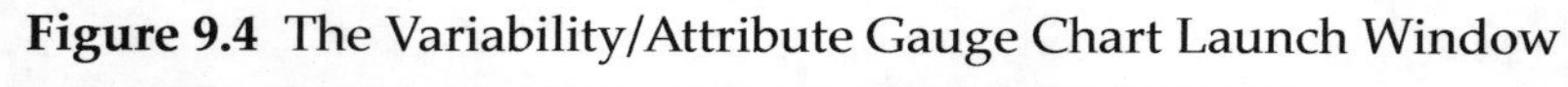

Figure 9.4 The Variability/Attribute Gauge Chart Launch Window

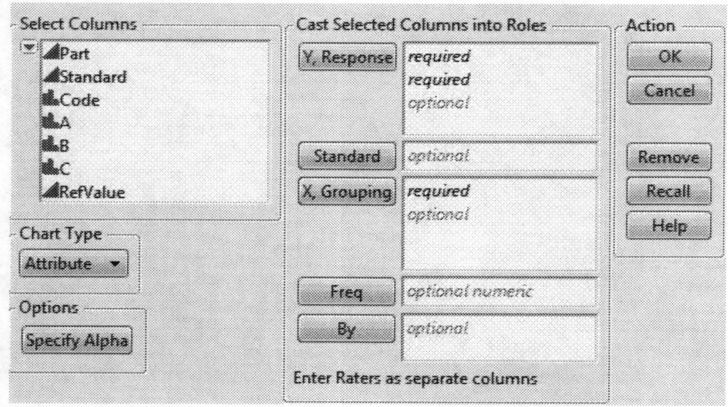

Chart Type Choose between a variability gauge analysis (for a continuous response) or an attribute gauge analysis (for a categorical response, usually "pass" or "fail").

Note: The content in this chapter covers only the **Attribute** chart type. For details about the **Variability** chart type, see "Variability Gauge Charts" chapter on page 175.

Specify Alpha Specify the alpha level used by the platform.

Y, Response Specify the columns of ratings given by each rater. You must specify more than one rating column.

Standard Specify a standard or reference column that contains the "true" or known values for the part. In the report window, an Effectiveness Report and an additional section in the Agreement Comparisons report appear, which compare the raters with the standard.

X, Grouping Specify the classification columns that group the measurements. If the factors form a nested hierarchy, specify the higher terms first.

Freq Identifies the data table column whose values assign a frequency to each row. Can be useful when you have summarized data.

By Identifies a column that creates a report consisting of separate analyses for each level of the variable.

For more information about the launch window, see the Get Started chapter in the *Using JMP* book.

The Attribute Gauge Chart and Reports

Attribute gauge chart plots the % Agreement, which is a measurement of rater agreement for every part in the study. The agreement for each part is calculated by comparing the ratings for every pair of raters for all ratings of that part. See "Statistical Details for Attribute Gauge Charts" on page 207.

Follow the instructions in "Example of an Attribute Gauge Chart" on page 200 to produce the results shown in Figure 9.5.

Figure 9.5 Attribute Gauge Chart

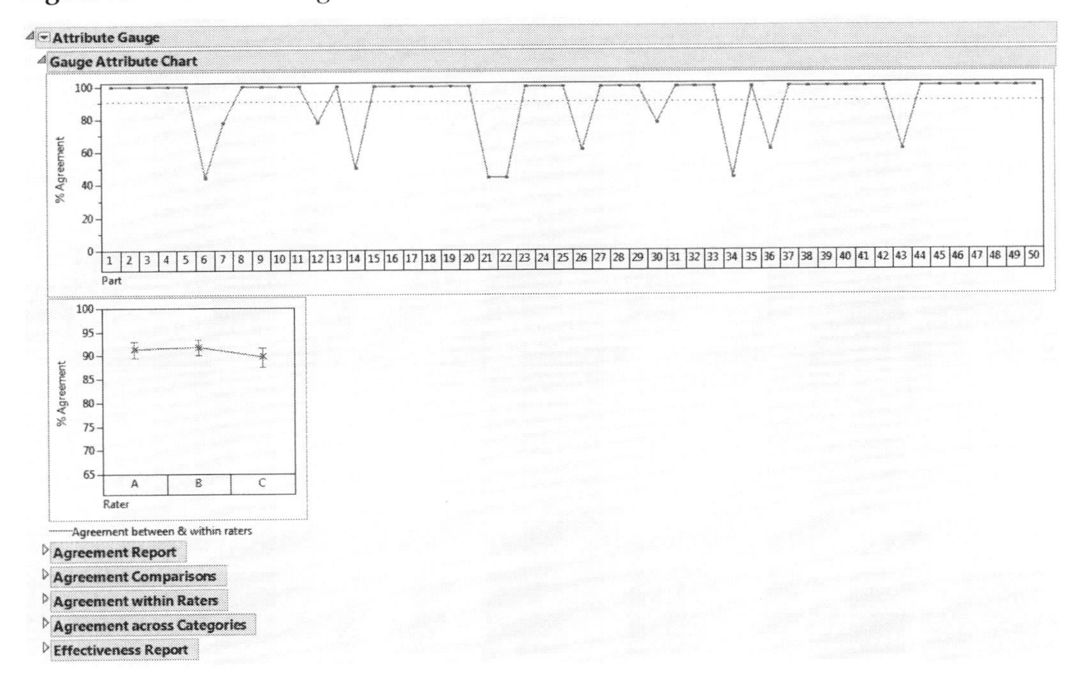

The first chart in Figure 9.5 uses all X grouping variables (in this case, the Part) on the *x*-axis. The second chart uses all Y variables on the *x*-axis (typically, and in this case, the Rater).

- In the first graph, you can look for parts with a low % Agreement value, and investigate to determine why raters do not agree about the measurement of that particular part.
- In the second graph, you can look for raters with a low % Agreement value, and investigate to determine why they do not agree with the other raters or with themselves.

For information about additional options, see "Attribute Gauge Platform Options" on page 206.

Agreement Reports

Note: The Kappa value is a statistic that expresses agreement. The closer the Kappa value is to 1, the more agreement there is. A Kappa value closer to 0 indicates less agreement.

The Agreement Report shows agreement summarized for each rater and overall agreement. This report is a numeric form of the data presented in the second chart in the Attribute Gauge Chart report. See Figure 9.5.

The Agreement Comparisons report shows each rater compared with all other raters, using Kappa statistics. The rater is compared with the standard only if you have specified a Standard variable in the launch window.

The Agreement within Raters report shows the number of items that were inspected. The confidence intervals are score confidence intervals (as suggested by Agresti and Coull, 1998). The Number Matched is the sum of the number of items inspected, where the rater agreed with him or herself on each inspection of an individual item. The Rater Score is the Number Matched divided by the Number Inspected.

The Agreement across Categories report shows the agreement in classification over that which would be expected by chance. It assesses the agreement between a fixed number of raters when classifying items.

Figure 9.6 Agreement Reports

Agreement Report

Rater	% Agreement	95% Lower CI	95% Upper CI
A	91.4286	89.5082	93.0248
B	91.9048	90.0502	93.4388
C	89.8095	87.6057	91.6588

Number Inspected	Number Matched	% Agreement	95% Lower CI	95% Upper CI
50	39	78.000	64.758	87.246

Agreement Comparisons

Rater	Compared with Rater	Kappa	Standard Error
A	B	0.8629	0.0442
A	C	0.7761	0.0547
B	C	0.7880	0.0537

Rater	Compared with Standard	Kappa	Standard Error
A	Standard	0.8788	0.0416
B	Standard	0.9230	0.0338
C	Standard	0.7740	0.0551

Agreement within Raters

Rater	Number Inspected	Number Matched	Rater Score	95% Lower CI	95% Upper CI
A	50	42	84.0000	71.4858	91.6626
B	50	45	90.0000	78.6398	95.6524
C	50	40	80.0000	66.9629	88.7562

Agreement across Categories

Category	Kappa	Standard Error
0	0.7936	0.0236
1	0.7936	0.0236
Overall	0.7936	0.0236

Effectiveness Report

The Effectiveness Report appears only if you have specified a Standard variable in the launch window. For a description of a Standard variable, see "Launch the Variability/Attribute Gauge Chart Platform" on page 201. This report compares every rater with the standard.

Figure 9.7 Effectiveness Report

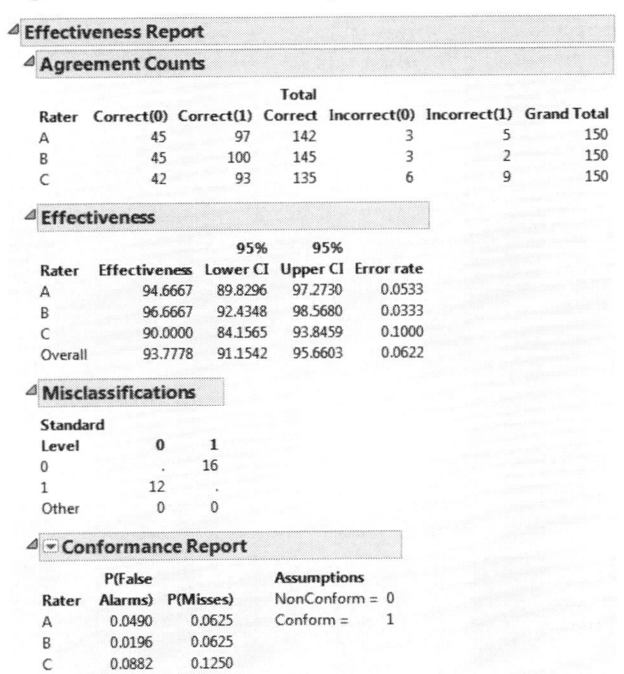

Effectiveness Report

Agreement Counts

Rater	Correct(0)	Correct(1)	Total Correct	Incorrect(0)	Incorrect(1)	Grand Total
A	45	97	142	3	5	150
B	45	100	145	3	2	150
C	42	93	135	6	9	150

Effectiveness

Rater	Effectiveness	95% Lower CI	95% Upper CI	Error rate
A	94.6667	89.8296	97.2730	0.0533
B	96.6667	92.4348	98.5680	0.0333
C	90.0000	84.1565	93.8459	0.1000
Overall	93.7778	91.1542	95.6603	0.0622

Misclassifications

Standard Level	0	1
0	.	16
1	12	.
Other	0	0

Conformance Report

Rater	P(False Alarms)	P(Misses)
A	0.0490	0.0625
B	0.0196	0.0625
C	0.0882	0.1250

Assumptions
NonConform = 0
Conform = 1

The Agreement Counts table shows cell counts on the number correct and incorrect for every level of the standard. In Figure 9.7, the standard variable has two levels, 0 and 1. Rater A had 45 correct responses and 3 incorrect responses for level 0, and 97 correct responses and 5 incorrect responses for level 1.

Effectiveness is defined as follows: the number of correct decisions divided by the total number of opportunities for a decision. For example, say that rater A sampled every part three times. On the sixth part, one of the decisions did not agree (for example, pass, pass, fail). The other two decisions would still be counted as correct decisions. This definition of effectiveness is different from the MSA 3rd edition. According to MSA, all three opportunities for rater A on part six would be counted as incorrect. Including all of the inspections separately gives you more information about the overall inspection process.

In the Effectiveness table, 95% confidence intervals are given about the effectiveness. These are score confidence intervals. It has been demonstrated that score confidence intervals provide increased coverage probability, particularly where observations lie near the boundaries. (See Agresti and Coull, 1998.)

The Misclassifications table shows the incorrect labeling. The rows represent the levels of the standard or accepted reference value. The columns contain the levels given by the raters.

Conformance Report

The Conformance Report shows the probability of false alarms and the probability of misses. The Conformance Report appears only when the rating has two levels (such as pass or fail, or 0 or 1).

The following descriptions apply:

False Alarm The part is determined to be non-conforming, when it actually is conforming.

Miss The part is determined to be conforming, when it actually is not conforming.

P(False Alarms) The number of parts that have been incorrectly judged to be nonconforming divided by the total number of parts that are judged to be conforming.

P(Miss) The number of parts that have been incorrectly judged to be conforming divided by the total number of parts that are actually nonconforming.

The Conformance Report red triangle menu contains the following options:

Change Conforming Category Reverses the response category that is considered conforming.

Calculate Escape Rate Calculates the Escape Rate, which is the probability that a non-conforming part is produced and not detected. The Escape Rate is calculated as the probability that the process will produce a non-conforming part times the probability of a miss. You specify the probability that the process will produce a non-conforming part, also called the Probability of Nonconformance.

Note: Missing values are treated as a separate category in this platform. To avoid this separate category, exclude rows of missing values in the data table.

Attribute Gauge Platform Options

The Attribute Gauge red triangle menu contains the following options:

Attribute Gauge Chart Shows or hides the gauge attribute chart and the efficiency chart.

Show Agreement Points Shows or hides the agreement points on the charts.

Connect Agreement Points Connects the agreement points in the charts.

Agreement by Rater Confid Intervals Shows or hides the agreement by rater confidence intervals on the efficiency chart.

Show Agreement Group Means Shows or hides the agreement group means on the gauge attribute chart. This option is available when you specify more than one X, Grouping variable.

Show Agreement Grand Mean Shows or hides the overall agreement mean on the gauge attribute chart.

Show Effectiveness Points Shows or hides the effectiveness points on the charts.

Connect Effectiveness Points Draws lines between the effectiveness points in the charts.

Effectiveness by Rater Confid Intervals Shows or hides confidence intervals on the second chart in the Attribute Gauge Chart report. See Figure 9.5.

Effectiveness Report Shows or hides the Effectiveness report. This report compares every rater with the standard, using the Kappa statistic.

See the JMP Reports chapter in the *Using JMP* book for more information about the following options:

Local Data Filter Shows or hides the local data filter that enables you to filter the data used in a specific report.

Redo Contains options that enable you to repeat or relaunch the analysis. In platforms that support the feature, the Automatic Recalc option immediately reflects the changes that you make to the data table in the corresponding report window.

Save Script Contains options that enable you to save a script that reproduces the report to several destinations.

Save By-Group Script Contains options that enable you to save a script that reproduces the platform report for all levels of a By variable to several destinations. Available only when a By variable is specified in the launch window.

Statistical Details for Attribute Gauge Charts

For the first chart in Figure 9.5 that plots all **X, Grouping** variables on the x-axis, the % Agreement is calculated as follows:

$$\%\ \text{Agreement for subject } i = \frac{\sum_{i=1}^{k} \binom{\text{number of responses for level } l}{2}}{\binom{N_i}{2}}$$

For the second chart in Figure 9.5 that plots all **Y, Response** variables on the *x*-axis, the % Agreement is calculated as follows:

$$\%\text{ Agreement for rater } k = \frac{\sum_{i=1}^{n}\left(\sum_{j=1}^{r_i} \begin{array}{c}\text{number of uncounted matching levels}\\ \text{for this rater k within part i for rep j}\end{array}\right)}{\sum_{i=1}^{n}\left(\sum_{j=1}^{r_i} N_i - j\right)}$$

Note the following:

- n = number of subjects (grouping variables)
- r_i = number of reps for subject i $(i = 1,...,n)$
- m = number of raters
- k = number of levels
- $N_i = m \times r_i$. Number of ratings on subject i $(i = 1,...,n)$. This includes responses for all raters, and repeat ratings on a part. For example, if subject i is measured 3 times by each of 3 raters, then N_i is 3 x 3 = 9.

For example, consider the following table of data for three raters, each having three replicates for one subject.

Table 9.1 Three Replicates for Raters A, B, and C

	A	B	C
1	1	1	1
2	1	1	0
3	0	0	0

Using this table, you can make these calculations:

$$\%\text{ Agreement} = \frac{\binom{4}{2} + \binom{5}{2}}{\binom{9}{2}} = \frac{16}{36} = 0.444$$

$$\%\text{ Agreement [rater A]} = \%\text{ Agreement [rater B]} = \frac{4+3+3}{8+7+6} = \frac{10}{21} = 0.476 \text{ and}$$

$$\%\text{ Agreement [rater C]} = \frac{4+3+2}{8+7+6} = \frac{9}{21} = 0.4286$$

Statistical Details for the Agreement Report

The simple Kappa coefficient is a measure of inter-rater agreement.

$$\hat{\kappa} = \frac{P_0 - P_e}{1 - P_e}$$

where:

$$P_0 = \sum_i p_{ii}$$

and:

$$P_e = \sum_i p_{i.}p_{.i}$$

If you view the two response variables as two independent ratings of the n subjects, the Kappa coefficient equals +1 when there is complete agreement of the raters. When the observed agreement exceeds chance agreement, the Kappa coefficient is positive, and its magnitude reflects the strength of agreement. Although unusual in practice, Kappa is negative when the observed agreement is less than the chance agreement. The minimum value of Kappa is between -1 and 0, depending on the marginal proportions.

Estimate the asymptotic variance of the simple Kappa coefficient with the following equation:

$$\text{var} = \frac{A + B - C}{(1 - P_e)^2 n}$$

where:

$$A = \sum_i p_{ii}\left[1 - (p_{i.} + p_{.i})(1 - \hat{\kappa})\right]$$

$$B = (1 - \hat{\kappa})^2 \sum_{i \neq j} \sum p_{ij}(p_{.i} + p_{j.})^2$$

and:

$$C = \left[\hat{\kappa} - P_e(1 - \hat{\kappa})\right]^2$$

The Kappas are plotted and the standard errors are also given.

Note: The Kappa statistics in the Attribute Chart platform are shown even when the levels of the variables are unbalanced.

Categorical Kappa statistics (Fleiss 1981) are found in the Agreement Across Categories report.

Given the following assumptions:

- n = number of subjects (grouping variables)
- m = number of raters
- k = number of levels
- r_i = number of reps for subject i ($i = 1,...,n$)
- $N_i = m \times r_i$. Number of ratings on subject i ($i = 1, 2,...,n$). This includes responses for all raters, and repeat ratings on a part. For example, if subject i is measured 3 times by each of 2 raters, then N_i is 3 x 2 = 6.
- x_{ij} = number of ratings on subject i ($i = 1, 2,...,n$) into level j (j=1, 2,...,k)The individual category Kappa is as follows:

$$\hat{\kappa}_j = 1 - \frac{\sum_{i=1}^{n} x_{ij}(N_i - x_{ij})}{(\bar{p}_j \bar{q}_j) \sum_{i=1}^{n} N_i(N_i - 1)} \quad \text{where} \quad \bar{p}_j = \frac{\sum_{i=1}^{n} x_{ij}}{\sum_{i=1}^{n} N_i} \quad \bar{q}_j = 1 - \bar{p}_j$$

The overall Kappa is as follows:

$$\bar{\hat{\kappa}} = \frac{\sum_{j=1}^{k} \bar{q}_j \bar{p}_j \hat{\kappa}_j}{\sum_{j=1}^{k} \bar{p}_j \bar{q}_j}$$

The variance of $\hat{\kappa}_j$ and $\bar{\hat{\kappa}}$ are as follows:

$$\mathrm{var}(\hat{\kappa}_j) = \frac{2}{nN(N-1)}$$

$$\mathrm{var}(\bar{\hat{\kappa}}) = \frac{2}{\left(\sum_{j=1}^{k} \bar{p}_j \bar{q}_j\right)^2 nN(N-1)} \times \left[\left(\sum_{j=1}^{k} \bar{p}_j \bar{q}_j\right)^2 - \sum_{j=1}^{k} \bar{p}_j \bar{q}_j (\bar{q}_j - \bar{p}_j)\right]$$

The standard errors of $\hat{\kappa}_j$ and $\bar{\hat{\kappa}}$ are shown only when there are an equal number of ratings per subject (for example, $N_i = N$ for all $i = 1, \ldots, n$).

Chapter 10

Process Capability

Measure the Variability of a Process over Time

Process capability analysis, used in process control, measures how well a process is performing compared to given specification limits. A good process is one that is stable and consistently produces product that is well within specification limits. A capability index is a measure that relates process performance, summarized by process centering and variability, to specification limits.

Graphical tools such as a goal plot and box plots give you quick visual ways of identifying which process or product characteristics are within specifications. Individual detail reports display a capability report for each variable in the analysis. The analysis enables you to identify variation relative to the specifications or requirements; this enables you to achieve increasingly higher conformance values.

You can specify subgroups to compare the overall variation of the process to the within subgroup variation. You can compute capability indices for processes that produce measurements that follow various distributions. For data that follow none of the specified distributions, you can compute nonparametric capability indices.

Figure 10.1 Example of the Process Capability Platform

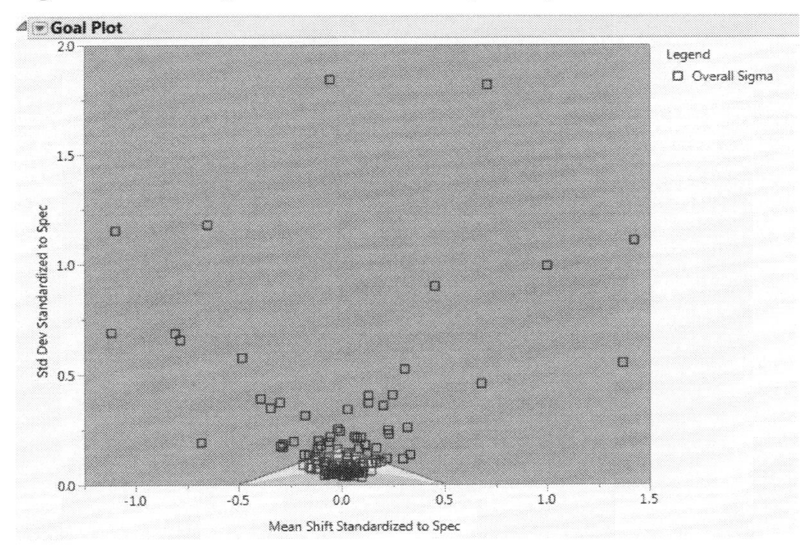

Process Capability Platform Overview

The Process Capability platform provides the tools needed to measure the compliance of a process to given specifications. By default, JMP shows a Goal Plot, Capability Box Plots, and a Capability Index Plot for the variables that you fit with normal distributions. Capability indices for nonnormal variables are plotted on the Capability Index Plot. You can add normalized box plots, summary reports, and individual detail reports for the variables in your analysis.

You can supply specification limits in several ways:

- in the data table, using a column property
- by requesting the Spec Limits Dialog in the launch window
- by loading the limits from a specification limits data table

You can specify two-sided, one-sided, or asymmetric specification limits.

Note: The Process Capability platform expands significantly on the Capability analyses that are available through Analyze > Distribution and through Analyze > Quality and Process > Control Chart.

Capability Indices

A capability index is a ratio that relates the ability of a process to produce product that meets specification limits. The index relates estimates of the mean and standard deviation of the quality characteristic to the specification limits. Within estimates of capability are based on an estimate of the standard deviation constructed from within-subgroup variation. Overall estimates of capability use an estimate of standard deviation constructed from all of the process data. See "Capability Indices for Normal Distributions" on page 267 and "Variation Statistics" on page 262.

Estimates of the mean or standard deviation are well-defined only if the processes related to centering or spread are *stable*. Therefore, interpretation of within capability indices requires that process spread is stable. Interpretation of overall capability indices requires that both process centering and spread are stable.

Capability indices constructed from small samples can be highly variable. The Process Capability platform provides confidence intervals for most capability indices. Use these to determine the range of potential values for your quality characteristic's actual capability.

Note: JMP PRO When confidence intervals are not provided (for example, for nonnormal distributions) you can use the Simulate feature to construct confidence intervals. For an example, see "Simulation of Confidence Limits for a Nonnormal Process Ppk" on page 256.

Guidelines for values of capability indices can be found in Montgomery (2013). The minimum recommended value is 1.33. Six Sigma initiatives aim for much higher capability levels that correspond to extremely low rates of defective parts per million.

Capability Indices for Nonnormal Processes

The Process Capability platform constructs capability indices for process measurements with the following distributions: Normal, gamma, Johnson, lognormal, and Weibull. A Best Fit option determines the best fit among these distributions and provides capability indices for this fit. The platform also provides a Nonparametric fit option that gives nonparametric estimates of capability.

For the nonnormal methods, estimates are constructed using two approaches: the ISO/Quantile method (Percentiles) and the Bothe/Z-scores method (Z-Score). For details about these methods, see "Capability Indices for Nonnormal Distributions: Percentile and Z-Score Methods" on page 267.

Note: Process Capability analysis for individual responses is accessible through Analyze > Quality and Process > Control Chart Builder. However, nonnormal distributions are available only in the Process Capability platform.

Overall and Within Estimates of Sigma

Most capability indices in the Process Capability platform can be computed based on estimates of the *overall* (long-term) variation and the *within*-subgroup (short-term) variation. If the process is stable, these two measures of variation should yield similar results since the overall and within subgroup variation should be similar. The normalized box plots and summary tables can be calculated using either the overall or the within-subgroup variation. See "Additional Examples of the Process Capability Platform" on page 249 for examples of capability indices computed for stable and unstable processes.

You can specify subgroups for estimating within-subgroup variation in the launch window. You can specify a column that defines subgroups or you can select a constant subgroup size. For each of these methods, you can choose to estimate the process variation using the average of the unbiased standard deviations or using the average of the ranges. If you do not specify subgroups, the Process Capability platform constructs a within-subgroup estimate of the process variation using a moving range of subgroups of size two. Finally, you can specify a historical sigma to be used as an estimate of the process standard deviation.

Capability Index Notation

The Process Capability platform provides two sets of capability indices. See "Capability Indices for Normal Distributions" on page 267 for details about the calculation of the capability indices.

- Cpk, Cpl, Cpu, Cp, and Cpm. These indices are based on a within-subgroup (short-term) estimate of the process standard deviation.
- Ppk, Ppl, Ppu, Pp, and Cpm. These indices are based on an overall (long-term) estimate of the process standard deviation. Note that the process standard deviation does not exist if the process is not stable. See Montgomery (2013).

The Process Capability platform uses the appropriate AIAG notation for capability indices: Ppk labeling denotes an index constructed from an overall variation estimate and Cpk denotes an index constructed from a within-subgroup variation estimate.

Note: The AIAG (Ppk) Labeling platform preference is selected by default. You can change the reporting to use Cp notation only by deselecting this preference under Process Capability.

For more information about process capability analysis, see Montgomery (2013) and Wheeler (2004).

Example of the Process Capability Platform with Normal Variables

This example uses the Semiconductor Capability.jmp sample data table. The variables represent standard measurements that a semiconductor manufacturer might make on a wafer as it is being processed. Specification limits for the variables have been entered in the data table through the Column Properties > Spec Limits property.

1. Select **Help > Sample Data Library** and open Semiconductor Capability.jmp.
2. Select **Analyze > Quality and Process > Process Capability**.
3. Select PNP1, PNP2, NPN2, PNP3, IVP1, PNP4, NPN3, and IVP2, and click **Y, Process**.
4. Click **OK**.
5. Select **Label Overall Sigma Points** from the Goal Plot red triangle menu.

Figure 10.2 Example Results for Semiconductor Capability.jmp

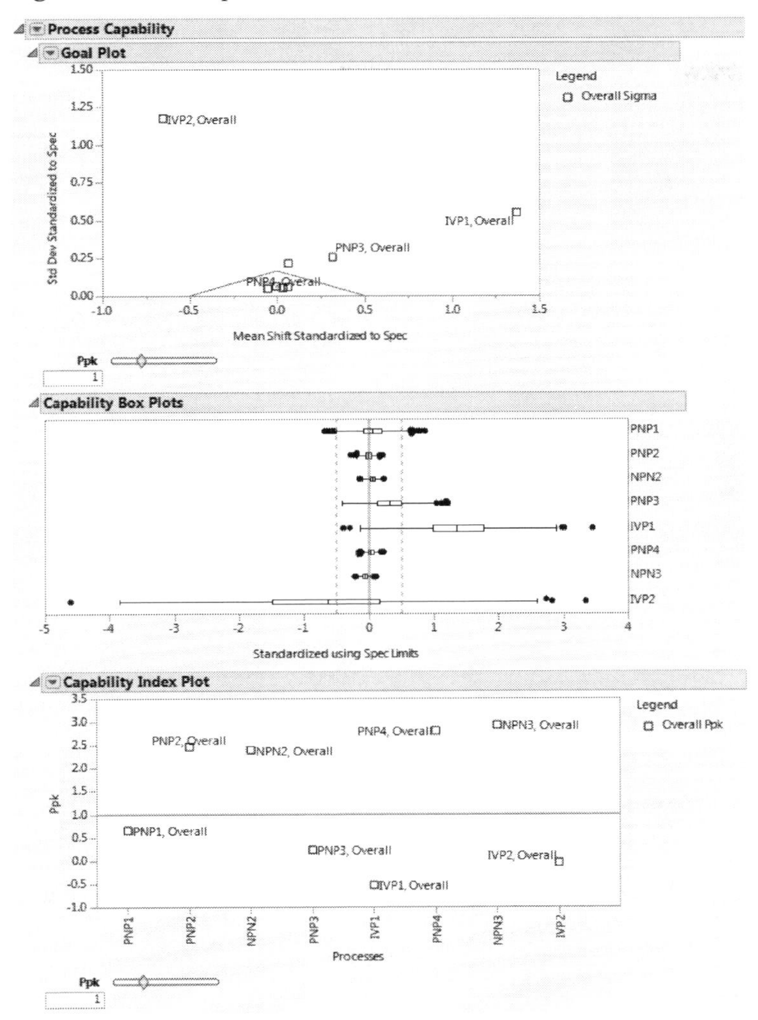

The Goal Plot in Figure 10.2 shows the spec-normalized mean shift on the x-axis and the spec-normalized standard deviation on the y-axis for each variable. The triangular region defined by the red lines in the bottom center of the plot is the goal triangle. It defines a region of capability index values. You can adjust the goal triangle using the Ppk slider below the plot. When the slider is set to 1, note that PNP1, PNP3, IVP1, and IVP2 are outside of the goal triangle and possibly out of specification.

The Capability Box Plots report shows a box plot for each variable in the analysis. The values for each column are centered by their target value and scaled by the specification range. In this example, all process variables have both upper and lower specification limits, and these are symmetric about the target value. It follows that the solid green line shows where the target should be and the dashed lines represent the specification limits.

It appears that the majority of points for IVP1 are above its upper specification limit (USL), and the majority of points for IVP2 are less than its target. PNP2 seems to be on target with all data values inside the specification limits.

The Capability Index Plot plots the Ppk values for each variable. Four variables come from very capable processes, with Ppk values of 2 or more. Four variables have Ppk values below 1.

Example of the Process Capability Platform with Nonnormal Variables

The Process Measurements.jmp data table contains measurements made on seven different processes used to construct a product. For each process, specification limits are saved as column properties. You begin by examining the distributions of your process data. You see that the distributions are not normal. Then you use the nonnormal capability features of the Process Capability platform to compute capability indices.

View the Distributions

1. Select **Help > Sample Data Library** and open Process Measurements.jmp.
2. Select **Analyze > Distribution.**
3. Select all seven columns from the **Select Columns** list and click **Y, Columns.**
4. Check the box next to **Histograms Only.**
5. Click **OK.**

   For most processes, the histograms show evidence that the theoretical distribution of measurements is skewed and does not follow a normal distribution. Therefore, for each process, you find the best fitting distributions among all of the available parametric distributions.

Perform a Capability Analysis

1. Select **Analyze > Quality and Process > Process Capability.**
2. Select all seven columns from the **Columns** list and click **Y, Process.**
3. Select all seven columns in the **Y, Process** list.
4. Open the **Distribution Options** panel and select **Best Fit** from the **Distribution** list.
5. Click **Set Process Distribution.**

   The suffix **&Dist(Best Fit)** is added to each variable name in the Y, Process list. The Best Fit option specifies that the best-fitting parametric distribution should be fit to each variable.

The available parametric distributions are normal, gamma, Johnson, lognormal, and Weibull. See Figure 10.3.

6. Open the **Nonnormal Distribution Options** outline. Note that the Nonnormal Capability Indices Method is set to **Percentiles**, the Johnson Distribution Fitting Method is set to **Quantile Matching**, and the Distribution Comparison Criterion is set to **AICc.**

Figure 10.3 Completed Launch Window

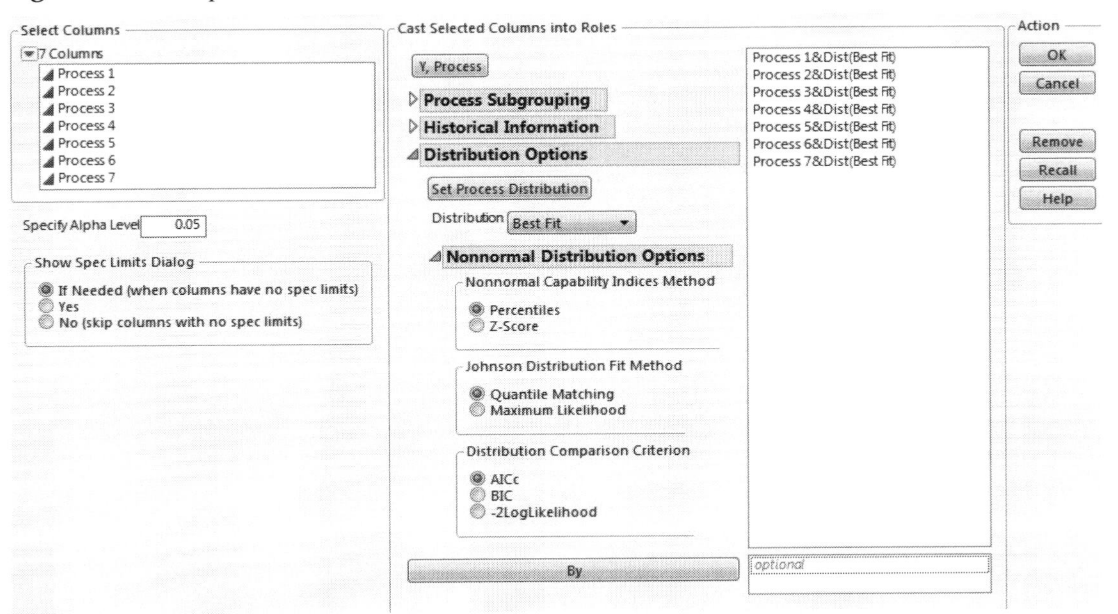

The Quantile Matching method is the default method used for fitting Johnson distributions because of its stability and speed as compared to Maximum Likelihood. Note that Maximum Likelihood is used in the Distribution platform.

7. Click **OK**.
8. Select **Label Overall Sigma Points** from the **Goal Plot** red triangle menu.
9. Select **Label Overall Sigma Points** from the Capability Index Plot red triangle menu.

Figure 10.4 Initial Report with Variables Labeled

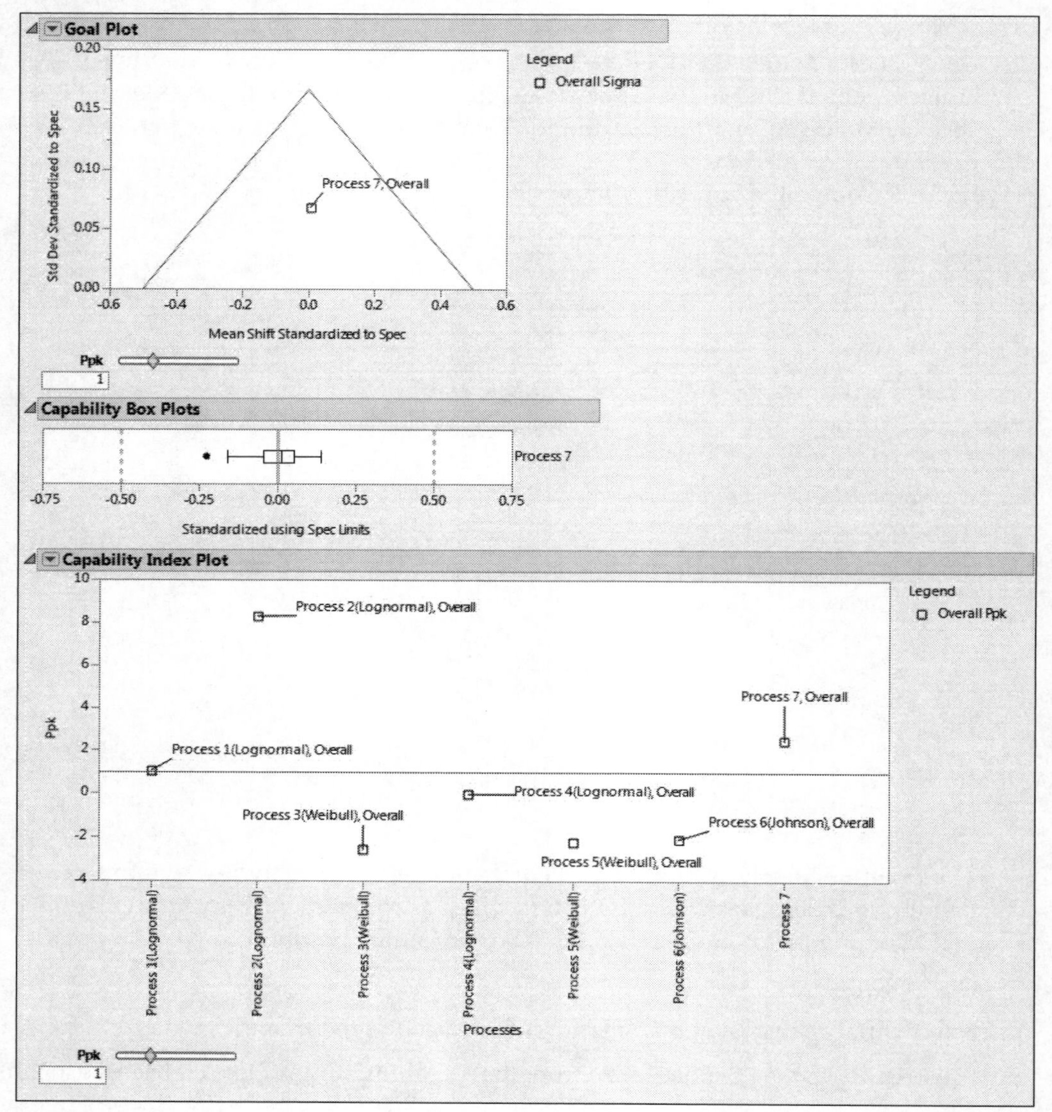

Note: Click on a label in the plot and drag it to make the plot more interpretable. Click on the right side frame of the Capability Index Plot and drag it to the right to make the labels easier to distinguish.

The Goal Plot shows only one point and it corresponds to Process 7. The Capability Box Plots report shows a single box plot for Process 7. This is because the best fit for Process 7 is a normal distribution.

10. Beneath the Capability Index Plot, set the Ppk value to 2.

    The Capability Index Plot shows Ppk values for all seven processes. Only two processes, Process 2 and Process 7, have capability values that exceed 2. Note that the best fitting nonnormal distributions are shown in parentheses to the right of the variable names in the Capability Index Plot. The best fitting distribution for Process 7 is not shown because it is a normal distribution.

11. Select **Individual Detail Reports** from the Process Capability red triangle menu.

    Because you requested Best Fit in the launch window, the Compare Distributions option has been selected from each distribution's red triangle menu.

12. Scroll to the report entitled **Process 4(Lognormal) Capability**.

Figure 10.5 Individual Detail Report for Process 4

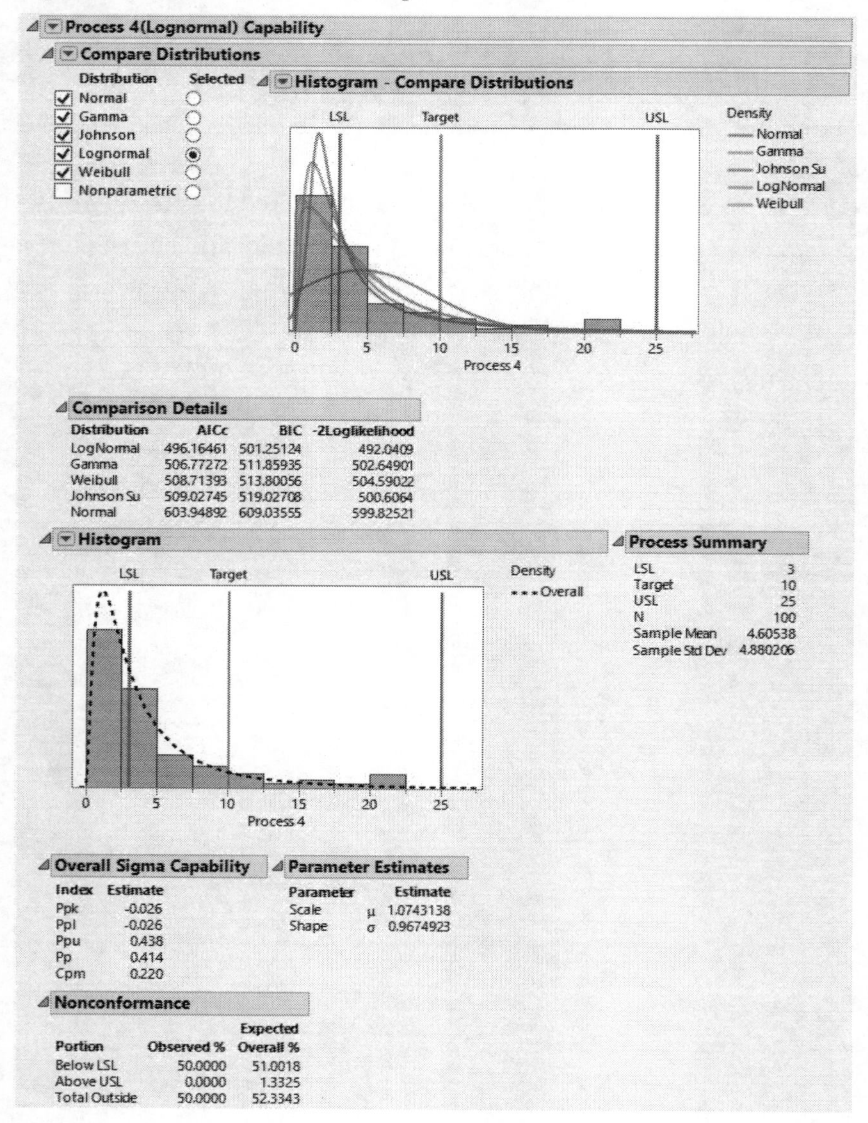

The title of the report for Process 4 indicates that the capability calculations are based on a lognormal fit. All of the check boxes in the Compare Distributions report, except the one for Nonparametric, are checked, indicating that these five distributions are fit. (This is because you requested a Best Fit in the launch window.) The button that is selected in the Selected column indicates that the Lognormal distribution is the distribution that is used in the remainder of the Process 4(Lognormal) Capability report to estimate capability and nonconformance.

The Compare Distributions report enables you to compare the five distributional fits. The Histogram - Compare Distributions report gives a visual assessment of the fit and the Comparison Details report shows fit statistics for the selected distributions. Both the plot and the fit statistics indicate that the lognormal distribution gives the best fit among the selected distributions.

The Individual Detail Report information that is shown by default includes a histogram showing the estimated best-fit distribution, a summary of the process information, capability indices based on an overall estimate of sigma, parameter estimates for the fitted lognormal distribution, and observed and expected nonconformance levels.

Launch the Process Capability Platform

Launch the Process Capability Platform by selecting Analyze > Quality and Process > Process Capability. In Figure 10.6, which uses the Semiconductor Capability.jmp data table, all outlines and panels have been opened.

Figure 10.6 Process Capability Launch Window

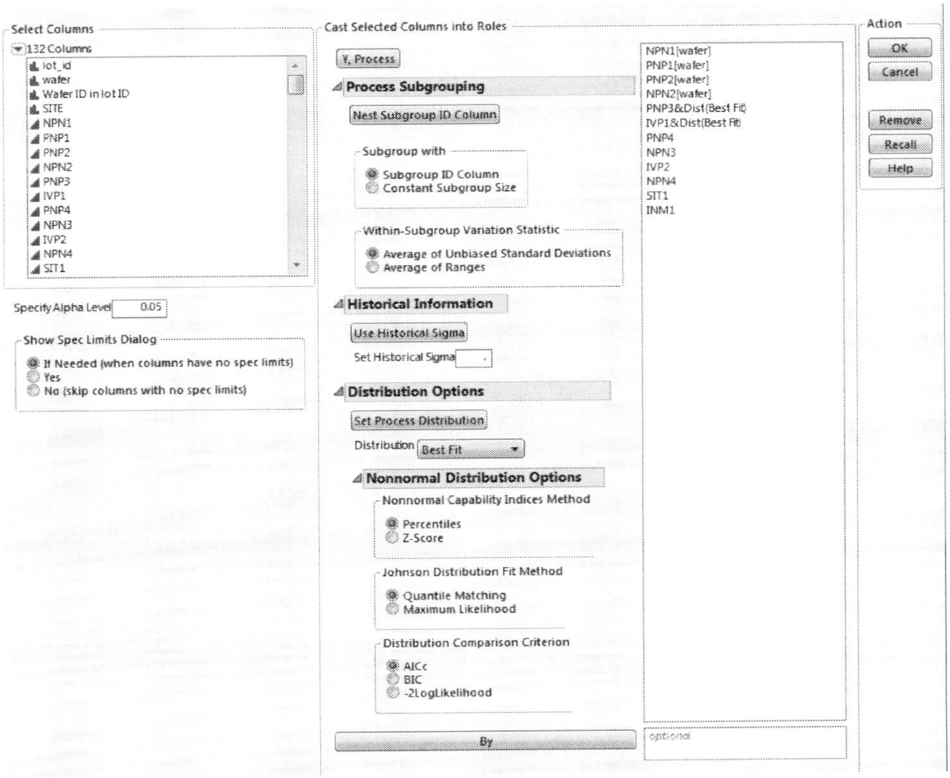

The Process Capability launch window contains the following outlines and options:

- "Process Selection" on page 224
- "Process Subgrouping" on page 224
- "Historical Information" on page 225
- "Distribution Options" on page 226
- "Other Specifications" on page 227

After you click OK in the launch window, the Spec Limits window appears unless one of the following occurs:

- All of the columns contain specification limits.
- You selected **No (skip columns with no spec limits)** on the launch window.

The Spec Limits window also appears if you select Yes on the launch window. Otherwise, the Process Capability report window appears.

Process Selection

Select the process variables to include in the capability analysis.

Y, Process Assigns the variables that you want to analyze.

Notes:

- The Transform menu is not available for the Select Column list in the Process Capability launch window. Right-click a column heading in the data table and select **New Formula Column** to create a transform column for use in Process Capability. See the Enter and Edit Data chapter in the *Using JMP* book for more information about creating new formula columns.
- Reference columns for virtually joined tables are not available in the Process Capability platform.

Process Subgrouping

This group of options enables you to assign each variable in the Y, Process list a subgroup ID column or a constant subgroup size.

Create Subgroups Using an ID Column

1. Select a variable or variables in the Y, Process list.
2. Select **Subgroup ID Column** from the **Subgroup with** options.
3. Select a subgroup ID column in the Select Columns list.
4. Click **Nest Subgroup ID Column**.

The subgroup ID column appears in brackets to the right of the variable names in the Y, Process list.

Create Subgroups Using a Constant Subgroup Size

1. Select a variable or variables in the Y, Process list.
2. Select **Constant Subgroup Size** from the **Subgroup with** options.
3. Enter the subgroup size next to **Set Constant Subgroup Size**.
4. Click **Subgroup by Size**.

The subgroup size appears in brackets to the right of the variable names in the Y, Process list.

Nest Subgroup ID Column Available when you select Subgroup ID Column. Assigns a column that you select from the Select Columns list to define the subgroups for the selected Y, Process columns.

Subgroup by Size Available when you select Constant Subgroup Size. Assigns the subgroup size that you specify in the Set Constant Subgroup Size box to define the subgroups for the selected Y, Process columns.

Set Constant Subgroup Size Available when you select Constant Subgroup Size. Specify the constant subgroup size for the selected Y, Process columns. You need to assign this value using Subgroup by Size.

Within-Subgroup Variation Statistic (Available when Process Subgrouping is used. Specifies if the within-subgroup estimate of standard deviation is calculated using standard deviations or ranges.

Historical Information

Use this outline to assign historically accepted values of the standard deviation to variables in the Y, Process list.

1. Select a variable or variables in the Y, Process list.
2. Enter a value next to Set Historical Sigma.
3. Select Use Historical Sigma to assign that value to the selected variables.

The specified value appears in parentheses in the expression "&Sigma()" to the right of the variable names in the Y, Process list.

Note: If you set a historical sigma, then subgroup assignments for the selected process variable are no longer relevant and are removed.

Distribution Options

Unless otherwise specified, all Y, Process variables are analyzed using the assumption that they follow a normal distribution. Use the Distribution Options outline to assign other distributions or calculation methods to variables in the Y, Process list and to specify options related to nonnormal calculations.

- The available distributions are the Normal, gamma, Johnson, lognormal, and Weibull distributions. Except for Johnson distributions, maximum likelihood estimation is used to fit distributions. See "Johnson Distribution Fit Method" on page 226.
- The Best Fit option determines the best fit among the available distributions and applies this fit.
- The Nonparametric option fits a distribution using kernel density estimation.

For more options related to nonnormal fits, see "Nonnormal Distribution Options" on page 226.

Specify a Distribution

1. Select a variable or variables in the Y, Process list.
2. Select a distribution from the Distribution list.
3. Select Set Process Distribution to assign that distribution to the selected variables.

The specified distribution appears in parentheses in the expression "&Dist()" to the right of the variable names in the Y, Process list.

Note: If you select a distribution other than Normal, you cannot assign a Subgroup ID column or a Historical Sigma. These selections are not supported by the methods used to calculate nonnormal capability indices. See "Capability Indices for Nonnormal Distributions: Percentile and Z-Score Methods" on page 267.

Nonnormal Distribution Options

Nonnormal Capability Indices Method Specifies the method used to compute capability indices for nonnormal distributions. See "Capability Indices for Nonnormal Distributions: Percentile and Z-Score Methods" on page 267.

Johnson Distribution Fit Method Specifies the method used to find the best-fitting Johnson distribution. Before estimating the parameters, the best-fitting family of distributions is determined from among the Johnson Su, Sb, and Sl families. The procedure described in Slifker and Shapiro (1980) is used to find the best-fitting family.

Quantile Matching The default method. It is more stable and faster than Maximum Likelihood. Quantile Matching Parameter estimates, assuming the best-fitting family, are obtained using a quantile-matching approach. See Slifker and Shapiro (1980).

Maximum Likelihood Parameters for the best-fitting family are determined using maximum likelihood.

Distribution Comparison Criterion (Available when a Best Fit Distribution is selected.) Specify the criterion that you want to use in determining a Best Fit. This criterion also determines the ordering of distributions in the Comparison Details report. See "Order by Comparison Criterion" on page 246.

Other Specifications

By Produces a separate report for each level of the By variable. If more than one By variable is assigned, a separate report is produced for each possible combination of the levels of the By variables.

Specify Alpha Level Specifies the significance level for confidence limits.

Show Spec Limits Dialog Specifies how to handle columns that do not have specification limits.

Note: It is good practice to ensure that specification limits for all process variables are specified as Spec Limits column properties or to load specification limits from a Limits Data table (see "Limits Data Table" on page 228). Otherwise, you can specify limits interactively in the Spec Limits window that appears after you click OK in the launch window (unless you select **No (skip columns with no spec limits)** on the launch window).

Entering Specification Limits

The lower specification limit (LSL), upper specification limit (USL), and target define the lower bound, upper bound, and target value for a quality process.

There are several ways to enter specification limits:

- Enter limits in the Spec Limits window after selecting columns in the launch window. See "Spec Limits Window" on page 228.
- Import limits from a JMP data table (known as a Limits Table). See "Limits Data Table" on page 228.
- Enter limits as Spec Limits column properties in the data table. See "Spec Limits Column Property" on page 229.
- If you are creating a Process Capability report by running a JSL script, enter limits in the script. See "The Process Capability Report" on page 230.

Only one specification limit is required for a selected column. If only the USL is specified, the box plots and Goal Plot point are colored blue. If only the LSL is specified, the box plots and Goal Plot point are colored red.

Spec Limits Window

After you click OK on the launch window, the Spec Limits window appears if any of the columns do not contain limits and you did not select **No (skip columns with no spec limits)** on the launch window. The Spec Limits window also appears if you select Yes on the launch window. Figure 10.7 shows the Spec Limits window for the Cities.jmp sample data table after selecting OZONE, CO, SO2, and NO as process variables in the launch window. Enter the known specification limits and click OK to view the Process Capability report.

Figure 10.7 Spec Limits Window for Cities.jmp

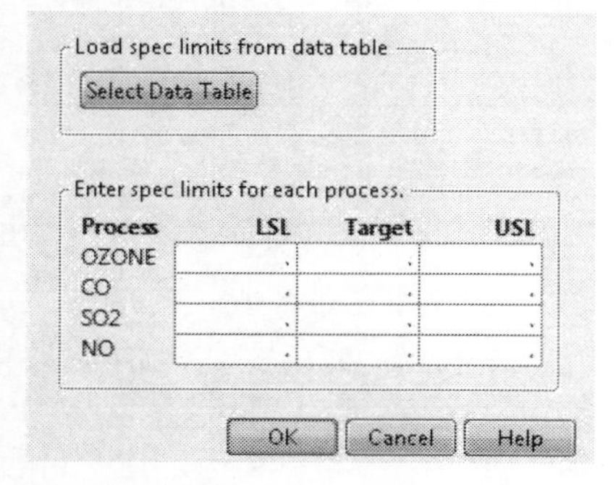

Limits Data Table

You can also specify a limits data table with the **Load spec limits from data table** option from the Spec Limits window. Click the **Select Data Table** button and then select the appropriate data table that contains the specification limits for the analysis. After you select the appropriate limits table, the values populate the window. Click **OK** to view the Process Capability report.

A limits data table can be in two different formats: *tall* or *wide*. A tall limits data table has one column for the responses and the limits key words are the other columns. A wide limits data table has a column for each response with one column to label the limits keys. Either of these formats can be read using the **Load spec limits from data table** option.

- A tall table contains four columns and has one row for each process. The first column has a character data type and contains the names of the columns analyzed in the Process Capability platform. The other three columns need to be named, _LSL, _USL, and _Target.

Figure 10.8 Example of a Tall Specification Limits Table

	Process	_LSL	_Target	_USL
1	OZONE	0.075	0.15	0.25
2	CO	5	7	12
3	SO2	0.01	0.04	0.09
4	NO	0.01	0.025	0.04

- A wide table contains three rows and one column for each column analyzed in the Process Capability platform plus a _LimitsKey column. In the _LimitsKey column, the three rows need to contain the identifiers _LSL, _USL, and _Target.

Figure 10.9 Example of a Wide Specification Limits Table

	_LimitsKey	OZONE	CO	SO2	NO
1	_LSL	0	5	0	0
2	_Target	0.05	10	0.03	0.025
3	_USL	0.1	20	0.08	0.6

The easiest way to create a limits data table is to save results computed by the Process Capability platform. The Save Spec Limits option in the Process Capability red triangle menu automatically saves limits from the sample values. After entering or loading the specification limits, you can do the following:

- Select **Save Spec Limits as Column Properties** to save the limits to the columns in the data table.
- Select **Save Spec Limits to New Table** to save the limits to a new tall specification limits data table. If you have selected at least one nonnormal distribution, a column called Distribution that contains the specified distributions is also added to the limits data table.

For more information, see "Process Capability Platform Options" on page 237.

Spec Limits Column Property

When you perform a capability analysis, you can use Column Properties > Spec Limits to save specification limits as a column property. The Spec Limits property applies only to numeric columns.

Some processes have one-sided specifications. Some have no target. You can enter any of these that apply: a lower specification limit, an upper specification limit, or a target value.

Figure 10.10 displays the Spec Limits section of the Column Properties window for OZONE in the sample data table Cities.jmp.

Figure 10.10 Spec Limits Section of the Column Properties Window

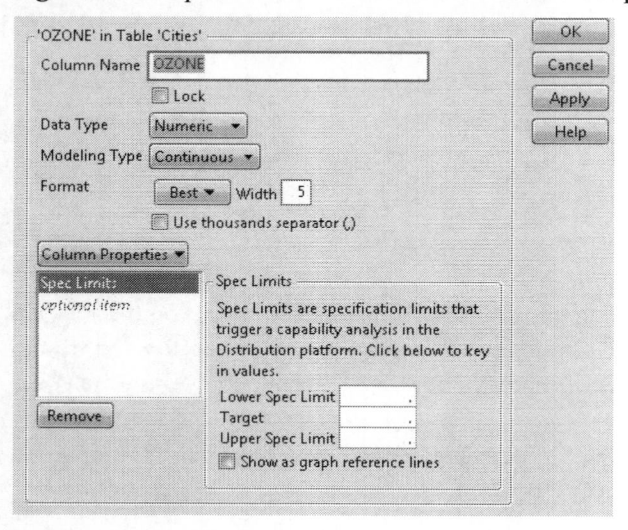

Tip: Saving specification limits as a column property ensures consistency when you repeat an analysis.

The Process Capability Report

By default, the Process Capability platform provides the following reports:

- "Goal Plot" on page 231 (provided only if at least one variable is fit with a normal distribution and shows only points for variables fit with normal distributions)
- "Capability Box Plots" on page 234 (provided only if at least one variable is fit with a normal distribution and shows only box plots for variables fit with normal distributions)
- "Capability Index Plot" on page 235

Figure 10.2 on page 217 shows an example of a default Process Capability report.

Using the Process Capability red triangle menu, you can add individual detail reports, normalized box plots, and summary reports. The red triangle menu also has options for identifying out-of-spec values in your data table, creating a summary data table, changing the display order of analyzed columns, and saving out spec limits. These options are described in "Process Capability Platform Options" on page 237.

You can change the default report at File > Preferences > Platforms > Process Capability. You can also make changes to the appearance of reports produced by options by selecting the relevant Process Capability topic at File > Preferences > Platforms.

Goal Plot

The Goal Plot shows, for each variable, the spec-normalized mean shift on the *x*-axis, and the spec-normalized standard deviation on the *y*-axis. It is useful for getting a quick, summary view of how the variables are conforming to specification limits. By default, the Goal Plot shows only those points for each column that are calculated using the overall sigma. Hold your cursor over each point to view the variable name and the sigma method used to calculate the point. See "Goal Plot" on page 264 for details about the calculation of the coordinates for the Goal Plot.

Note: Process variables with distributions other than Normal are not plotted on the Goal Plot.

Goal Plot Points

Points on the Goal Plot correspond to columns, not rows. Selecting a point in the Goal Plot selects the corresponding column in the data table.

The points on the Goal Plot are also linked to the rows of the Goal Plot Summary Table, where each row corresponds to a column. You can select a point in the Goal Plot, right-click, and apply row states. These row states are applied to the rows of the Goal Plot Summary Table. Row states that you apply in the Goal Plot Summary Table are reflected in the Goal Plot. To see this table, select Make Goal Plot Summary Table from the Process Capability red triangle menu. See "Make Goal Plot Summary Table" on page 248.

Tip: If you hide a point in the Goal Plot, you can show the point again by changing the corresponding row state in the Goal Plot Summary Table.

Goal Plot Triangle

The goal plot triangle appears in the center of the bottom of the Goal Plot. The slider beneath the plot enables you to adjust the size of goal triangle in the plot.

By default, the Ppk slider and the value beneath it are set to Ppk = 1. This approximates a non-conformance rate of 0.0027, if the distribution is normal. The goal triangle represents the Ppk shown in the box. To change the Ppk value, move the slider or enter a number in the box.

JMP gives the Goal Plot in terms of Ppk values by default. You can change this preference at File > Preferences > Platforms > Process Capability. When the AIAG (Ppk) Labeling preference is unchecked, all of the Ppk labeling is changed to Cpk labeling, including the label of the slider under the goal plot.

Goal Plot Options

The Goal Plot red triangle menu has the following options:

Show Within Sigma Points Shows or hides the points calculated using the within sigma estimate.

Show Overall Sigma Points Shows or hides the points calculated using the overall sigma estimate.

Shade Levels Shows or hides the Ppk level shading. See Figure 10.11. When you select Shade Levels, shaded areas appear in the plot. The shaded areas are described as follows, with *p* representing the value shown in the box beneath Ppk:

– Points in the red area have Ppk < *p*.

– Points in the yellow area have *p* < Ppk < 2*p*.

– Points in the green area have 2*p* < Ppk.

Label Within Sigma Points Shows or hides labels for points calculated using the within sigma estimate.

Label Overall Sigma Points Shows or hides labels for points calculated using the overall sigma estimate.

Defect Rate Contour Shows or hides a contour representing a specified defect rate.

Figure 10.11 shows the Goal Plot for the entire data set for the Semiconductor Capability.jmp sample data table after selecting Shade Levels and Show Within Sigma Points from the Goal Plot red triangle menu.

Figure 10.11 Goal Plot

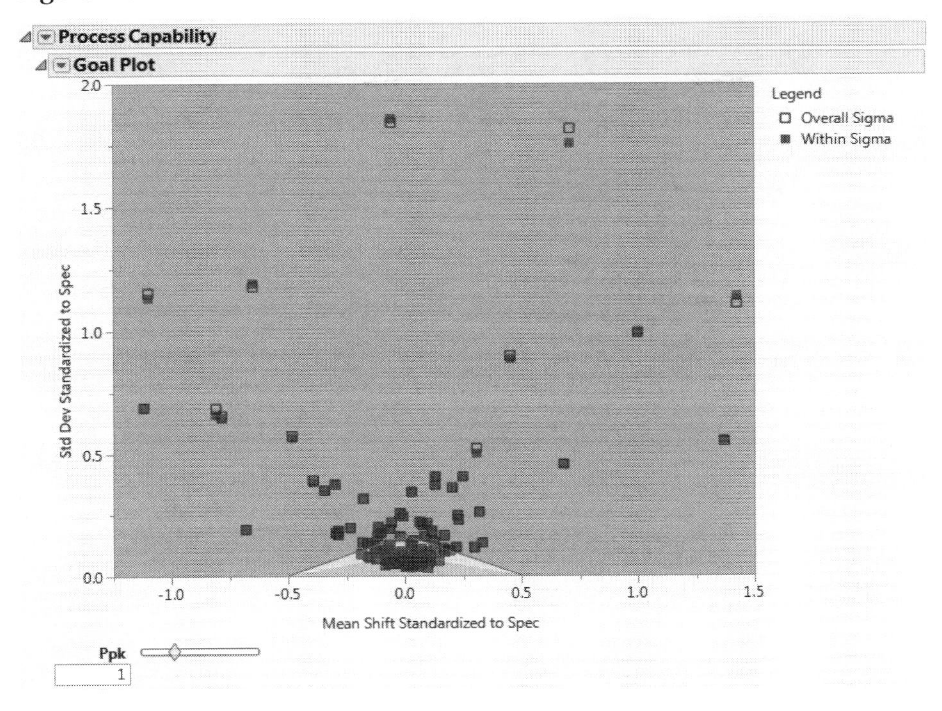

One-Sided or Missing Specification Limits

When there is only one specification limit for a column, markers and colors are used in the following ways:

- If only the upper specification limit (USL) is specified, the point on the Goal Plot is represented by a right-pointing triangle and is colored blue.
- If only the lower specification limit (LSL) is specified, the point on the Goal Plot is represented by a left-pointing triangle and is colored red.
- If at least one process has only an upper specification limit, the right half of the goal triangle is blue.
- If at least one process has only a lower specification limit, the left half of the goal triangle is red.

Processes with only an upper specification limit are represented by blue and should be compared to the blue (right) side of the goal triangle. Processes with only a lower specification limit are represented by red and should be compared to the red (left) side of the goal triangle. For details about how the coordinates of points are calculated, see "Goal Plot" on page 264.

Capability Box Plots

The Capability Box Plots show a box plot for each variable selected in the analysis. The values for each column are centered by their target value and scaled by the difference between the specification limits. If the target is not centered between the specification limits, the values are scaled by twice the minimum difference between the target and specification limits. For each process column Y_j (see "Notation for Goal Plots and Capability Box Plots" on page 264 for a description of the notation):

$$Z_{ij} = \frac{Y_{ij} - T_j}{2 \times \min(T_j - \mathrm{LSL}_j, \mathrm{USL}_j - T_j)}$$

For a process with a one-sided specification, see "One-Sided or Missing Specification Limits" on page 233. For the situation where no target is specified, see "Capability Box Plots for Processes with Missing Targets" on page 266.

Note: Process variables with distributions other than Normal are not plotted on the Capability Box Plot.

Figure 10.11 shows a Capability Box Plots report for eight variables in the Semiconductor Capability.jmp sample data table.

Figure 10.12 Capability Box Plot

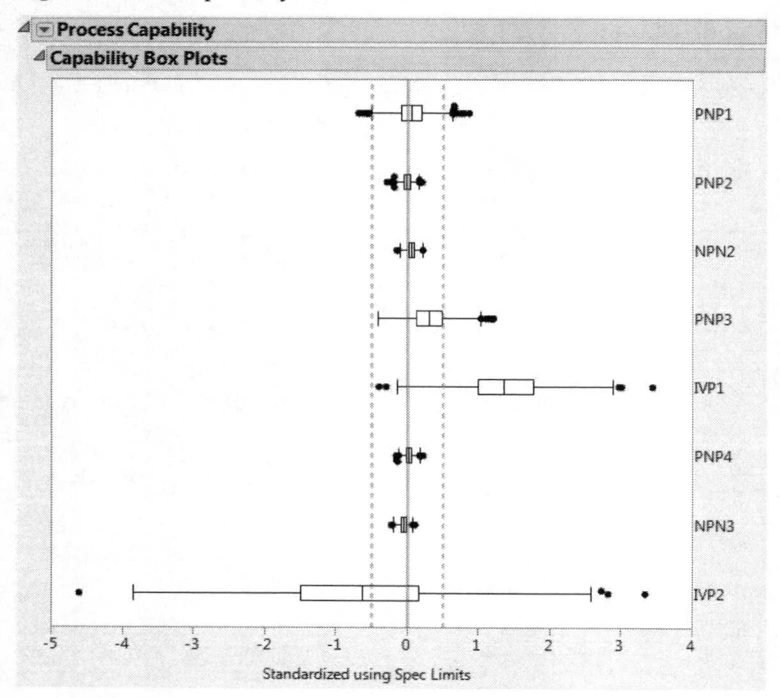

The plot displays dotted green lines drawn at ±0.5.

- For a process with a target that is centered between its specification limits, the dotted green lines represent the standardized specification limits.
- For a process with a target that is not centered between its specification limits, one of the dotted green lines represents the standardized specification limit for the limit closer to the target. The other dotted green line represents the same distance in the opposite direction.

This plot is useful for comparing variables with respect to their specification limits. For example, in Figure 10.12, the majority of points for IVP1 are above its USL, and the majority of its points for IVP2 are less than its target. PNP2 seems to be on target with all data points in the specification limits.

One-Sided or Missing Specification Limits

When there is only one specification limit for a column, colors are used in the following ways:

- If only the upper specification limit (USL) is specified, the box plot is colored blue.
- If only the lower specification limit (LSL) is specified, the box plot is colored red.
- If at least one process has only an upper specification limit, the dotted line at 0.5 is blue.
- If at least one process has only a lower specification limit, the dotted line at -0.5 is red.

Suppose that only the lower specification limit is specified and that the process target is specified. The capability box plot is based on the following values for the transformed observations. See "Notation for Goal Plots and Capability Box Plots" on page 264 for a description of the notation:

$$Z_{ij} = \frac{Y_{ij} - T_j}{2(T_j - \mathrm{LSL}_j)}$$

Suppose that only the upper specification limit is specified and that the process target is specified. The capability box plot is based on the following values for the transformed observations:

$$Z_{ij} = \frac{Y_{ij} - T_j}{2(\mathrm{USL}_j - T_j)}$$

For details about how missing targets are handled with one-sided specification limits, see "Single Specification Limit and No Target" on page 266.

Capability Index Plot

The Capability Index Plot shows Ppk values for all variables that you entered as Y, Process.

- Each variable name appears on the horizontal axis. If you fit a nonnormal distribution, the fitted distribution name appears in the plot as a parenthetical suffix to the variable name.
- The vertical axis shows Ppk values.
- A horizontal line is placed at the Ppk value specified by the slider beneath the plot.

Figure 10.13 shows a Capability Index Plot report for the **Process Measurements.jmp** sample data table. Seven of the variables are fit with nonnormal distributions. **Process 7** is fit with a normal distribution. Points have been labeled using the Label Overall Sigma Points option that is available in the Capability Index Plot red triangle menu.

Figure 10.13 Capability Index Plot with Nonnormal Distributions

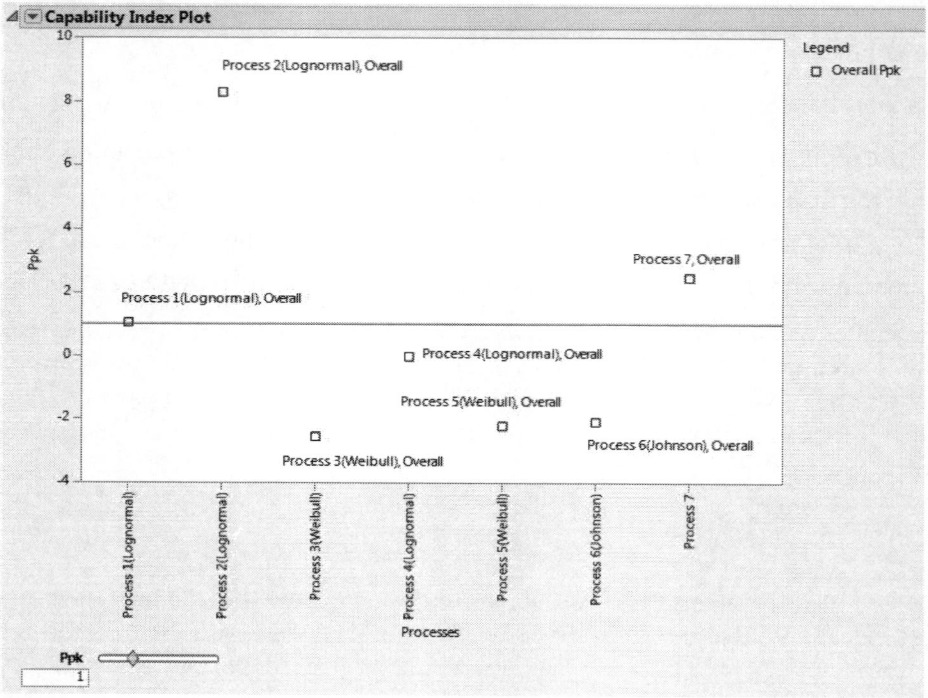

Capability Index Plot Options

The Capability Index Plot red triangle menu has the following options:

Show Within Sigma Points Shows or hides the points calculated using the within sigma estimate.

Show Overall Sigma Points Shows or hides the points calculated using the overall sigma estimate.

Shade Levels Shows or hides the Ppk level shading. When you select Shade Levels, shaded areas appear in the plot. The shaded areas are described as follows, with p representing the value shown in the box beneath Ppk:

– Points in the red area have Ppk < p.

– Points in the yellow area have p < Ppk < $2p$.

– Points in the green area have $2p$ < Ppk.

Label Within Sigma Points Shows or hides labels for points calculated using the within sigma estimate.

Label Overall Sigma Points Shows or hides labels for points calculated using the overall sigma estimate.

Process Capability Platform Options

The Process Capability red triangle menu contains the following options:

Individual Detail Reports Shows or hides individual detail reports for each variable in the analysis. See "Individual Detail Reports" on page 239 for more information.

Goal Plot Shows or hides a goal plot for the data. The Goal Plot shows the spec-normalized mean shift on the x-axis and the spec-normalized standard deviation on the y-axis for each variable. See "Goal Plot" on page 231 for more information. (Only variables for which you specify normal distributions are shown on the plot.)

Capability Box Plots Shows or hides a capability box plot for each variable in the analysis. The values for each column are centered by their target value and scaled by twice the minimum difference between the target value and the specification limits. See "Capability Box Plots" on page 234 for more information. (Box plots are shown only for variables for which you specify normal distributions.)

Normalized Box Plots Provides two options for plots that show normalized box plots for each process variable. Each column is standardized by subtracting its mean and dividing by an estimate of the column's standard deviation. The box plot is constructed using quantiles for the standardized values. See "Normalized Box Plots" on page 246 for more information. (Normalized box plots are shown only for variables for which you specify normal distributions.)

Within Sigma Normalized Box Plots Shows or hides a plot called Within Sigma Normalized Box Plots. The box plots are constructed using the within-subgroup estimate of standard deviation.

Overall Sigma Normalized Box Plots Shows or hides a plot called Overall Sigma Normalized Box Plots. The box plots are constructed using the overall estimate of standard deviation.

Capability Index Plot Shows overall Ppk values for all variables that you entered as Y, Process. See "Capability Index Plot" on page 235.

Summary Reports Provides two options for summary reports of capability indices. See "Summary Reports" on page 247 for more information.

Within Sigma Summary Report Shows or hides a summary report of capability indices calculated using the within-subgroup estimate of standard deviation. (Results are available only for variables with specified normal distributions.)

Overall Sigma Summary Report Shows or hides a summary report of capability indices calculated using the overall estimate of standard deviation.

Action Options

The following red triangle menu options perform actions:

Out of Spec Values Provides options for the cells in the data table containing values that are out of spec.

Select Out of Spec Values Selects all rows and columns in the data table that contain at least one value that does not fall within the specification limits.

Color Out of Spec Values Colors the cells in the data table that correspond to values that are out of spec. The cell is colored blue if the value is above the USL and red if the value is below the LSL.

Tip: To remove colors in specific cells, select all cells of interest. Right-click in one of the cells and select Clear Color. To remove colors in all cells, deselect Color Out of Spec Values.

Make Goal Plot Summary Table Creates a summary table for the points plotted in the Goal Plot. This table includes the variable's name, its spec-normalized mean shift, and its spec-normalized standard deviation. Each variable has two rows in this table: one for each sigma type (within and overall). See "Make Goal Plot Summary Table" on page 248 for more information.

Order By Reorders the box plots, summary reports, and individual detail reports. You can reorder by Initial Order, Reverse Initial Order, Within Sigma Cpk Ascending, Within Sigma Cpk Descending, Overall Sigma Ppk Ascending, or Overall Sigma Ppk Descending. The options that order by Within Sigma reorder plot elements only for variables with specified normal distributions.

Save Spec Limits Provides options for saving specification limits.

Save Spec Limits as Column Properties Saves the specification limits to a column property for each variable in the analysis. If no spec limit column property is present, the column property is created. If a spec limit column property is present, the values in the column property are overwritten. See "Spec Limits Column Property" on page 229 for more information.

Save Spec Limits to New Table Saves the specification limits to a limits data table in tall format. See "Limits Data Table" on page 228 for more information.

Save Distributions as Column Properties Saves the distribution used in calculating capability as a Process Capability Distribution column property. See the Column Info Window chapter in the *Using JMP* book.

If a column contains the Distribution property specifying a nonnormal distribution and no Process Capability Distribution property, then the Process Capability platform applies a nonnormal fit. The Process Capability platform uses the distribution specified in the Distribution column property, or a Johnson fit if that distribution is not supported in Process Capability. If a column contains the Process Capability Distribution property, then the Process Capability platform uses the distribution specified in the Process Capability Distribution column property.

Note: If you want to use a specific distribution in the Process Capability platform, save it as a Process Capability Distribution column property.

Relaunch Dialog Opens the platform launch window and recalls the settings used to create the report.

See the JMP Reports chapter in the *Using JMP* book for more information about the following options:

Local Data Filter Shows or hides the local data filter that enables you to filter the data used in a specific report.

Redo Contains options that enable you to repeat or relaunch the analysis. In platforms that support the feature, the Automatic Recalc option immediately reflects the changes that you make to the data table in the corresponding report window.

Save Script Contains options that enable you to save a script that reproduces the report to several destinations.

Save By-Group Script Contains options that enable you to save a script that reproduces the platform report for all levels of a By variable to several destinations. Available only when a By variable is specified in the launch window.

Individual Detail Reports

The Individual Detail Reports option displays a capability report for each variable in the analysis.

Normal Distributions

Figure 10.14 shows the Individual Detail Report for PNP1 from the Semiconductor Capability.jmp sample data table as described in "Example of the Process Capability Platform with Normal Variables" on page 216.

Figure 10.14 Individual Detail Report

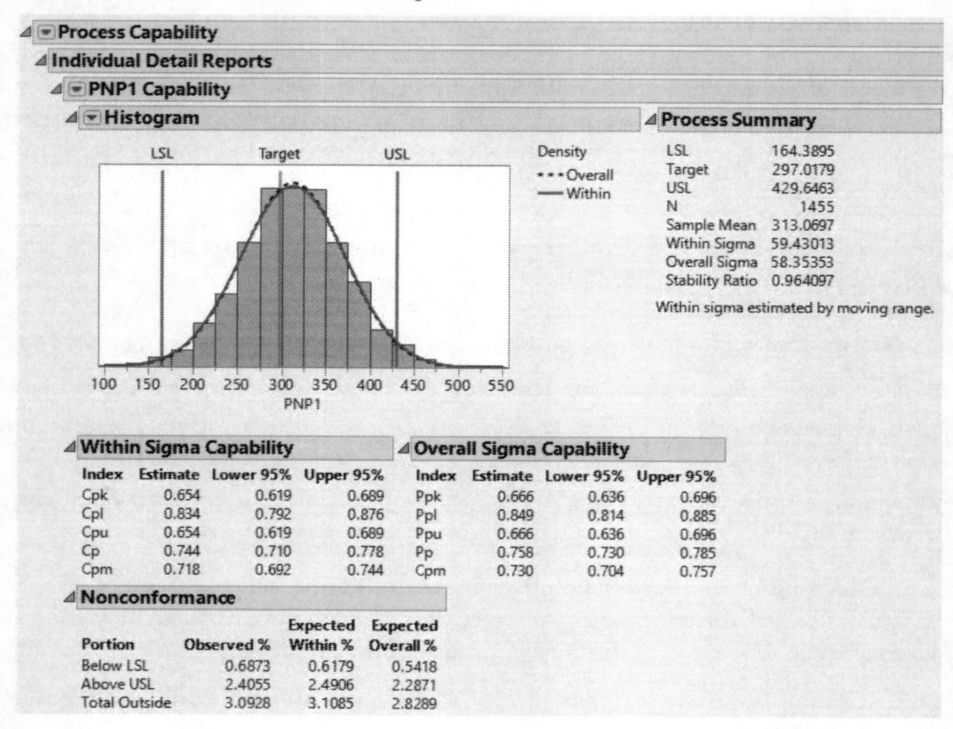

The Individual Details report for a variable with a normal distribution shows a histogram, process summary details, and capability and nonconformance statistics. The histogram shows the distribution of the values, the lower and upper specification limits and the process target (if they are specified), and one or two curves showing the assumed distribution. The histogram in Figure 10.14 shows two normal curves, one based on the overall estimate of standard deviation and the other based on the within-subgroup estimate.

When you fit your process with a normal distribution, the Process Summary includes the *Stability Ratio*, which is a measure of stability of the process. The stability ratio is defined as follows:

$$(\text{Overall Sigma}/\text{Within Sigma})^2$$

A stable process has stability ratio near one. Higher values indicate less stability.

Nonnormal Distributions

Note: Capability indices based on within-subgroup variation and stability ratios are not available for processes for which you have specified nonnormal distributions.

Figure 10.15 shows the Individual Detail Report for Process 1 from the Process Measurements.jmp sample data table as described in "Example of the Process Capability Platform with Nonnormal Variables" on page 218.

Figure 10.15 Individual Detail Report for Process 1

Individual Detail Reports
Johnson distributions fit with the Quantile Matching method.
Nonnormal capability indices calculated with the Percentiles method.

Process 1(Lognormal) Capability

Histogram

LSL Target USL
Density
Overall
0 5 10 15 20 25
Process 1

Process Summary

LSL	3
Target	10
USL	25
N	100
Sample Mean	10.07873
Sample Std Dev	3.158274

Overall Sigma Capability

Index	Estimate
Ppk	1.051
Ppl	1.139
Ppu	1.051
Pp	1.076
Cpm	1.024

Parameter Estimates

Parameter		Estimate
Scale	μ	2.2630009
Shape	σ	0.3085345

Nonconformance

Portion	Observed %	Expected Overall %
Below LSL	0.0000	0.0080
Above USL	0.0000	0.0974
Total Outside	0.0000	0.1054

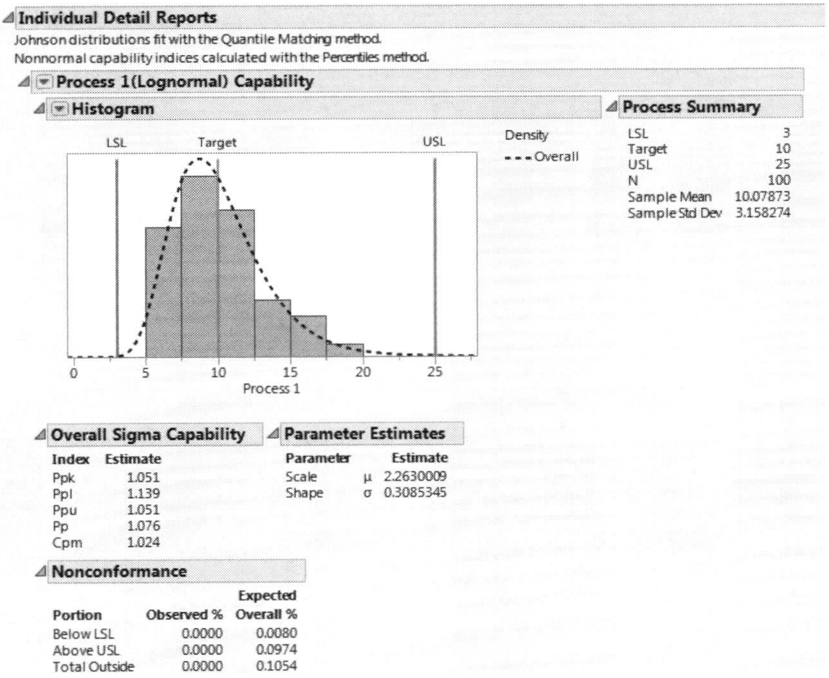

The report opens with a note summarizing the Nonnormal Distribution Options that you selected in the launch window.

The Individual Details report for a variable with a nonnormal distribution shows a histogram, process summary details, and capability and nonconformance statistics. The histogram shows the distribution of the values, the lower and upper specification limits and the process target (if they are specified). A curve showing the fitted distribution is superimposed on the histogram. If you selected a Nonparametric distribution, the curve shown in the histogram is the nonparametric density.

The report also shows a Parameter Estimates report if you selected a nonnormal parametric distribution or a Nonparametric Density report if you selected a Nonparametric fit. See "Parameter Estimates" on page 243 and "Nonparametric Density" on page 243.

Individual Detail Report Options

The outline title for each variable in the Individual Detail Reports section is of the form <Variable Name> Capability. However, if you request nonnormal capability, the relevant distribution name is shown parenthetically in the outline title.

Each Capability report has a red triangle menu with the following options:

Compare Distributions Shows or hides the control panel for comparing distributions for the process. See "Compare Distributions" on page 243.

Process Summary Shows or hides the summary statistics for the variable, including the overall sigma estimate, and, if you have specified a normal distribution, the within sigma estimate and the stability ratio.

Histogram Shows or hides the histogram of the values of the variable. The histogram report includes a red triangle menu that controls the following features of the histogram:

Show Spec Limits Shows or hides vertical red lines on the histogram at the specification limits for the process.

Show Target Shows or hides a vertical green line on the histogram at the process target.

Show Within Sigma Density Shows or hides an approximating normal density function on the histogram with mean given by the sample mean and standard deviation given by the within estimate of sigma.

Show Overall Sigma Density Shows or hides an approximating normal density function on the histogram with mean given by the sample mean and standard deviation given by the overall estimate of sigma.

Show Count Axis Shows or hides an additional axis to the right of the histogram plot showing the count of observations.

Show Density Axis Shows or hides an additional axis to the right of the histogram plot showing the proportion of observations.

Capability Indices Controls display of the following capability index reports:

Within Sigma Capability (Available when distribution is Normal.) Shows or hides capability indices (and confidence intervals) based on the within (short-term) sigma.

Within Sigma Z Benchmark (Available when distribution is Normal.) Shows or hides Z benchmark indices based on the within (short-term) sigma.

Overall Sigma Capability Shows or hides capability indices (and confidence intervals) based on the overall (long-term) sigma.

Overall Sigma Z Benchmark (Available when distribution is Normal.) Shows or hides Z benchmark indices based on the overall (long-term) sigma.

Nonconformance Shows or hides the observed and expected percentages of observations below the LSL, above the USL, and outside of the specification limits. The Nonconformance table contains hidden columns for observed and expected PPM and counts.

Parameter Estimates (Available when a distribution other than Normal or Nonparametric is selected.) Shows or hides the Parameter Estimates report, which gives estimates for the parameters of the selected distribution.

The estimates for all except the Johnson family distributions are obtained using maximum likelihood. For details about Johnson family fits, see "Johnson Distribution Fit Method" on page 226.

The parameters and probability density functions for the normal, gamma, Johnson, lognormal, and Weibull distributions are described in "Capability Indices for Nonnormal Distributions: Percentile and Z-Score Methods" on page 267. These are the same parameterizations used in the Distribution platform, with the exception that Process Capability does not support threshold parameters. See the Distributions chapter in the *Basic Analysis* book.

Nonparametric Density (Available when Nonparametric is selected as the distribution.) Shows or hides the Nonparametric Density report, which gives the *kernel bandwidth* used in fitting the nonparametric distribution. The kernel bandwidth is given by the following, where n is the number of observations and S is the uncorrected sample standard deviation:

$$\text{bandwith} = \frac{0.9S}{\sqrt[5]{n}}$$

Compare Distributions

The Compare Distributions report enables you to compare and apply various distributional fits. Note the following:

- Your selected distribution is indicated in the Selected column.
- The report initially shows fit statistics for your Selected distribution and other fitted distributions in the Comparison Details report. If you selected Best Fit, the Comparison Details report initially shows statistics for all parametric fits.
- Check the distributions in the Distribution list that you want to compare.
  - The probability density function for the best fitting distribution in each family that you select is superimposed on the histogram in the Histogram - Compare Distributions report.

- If the distribution is parametric, a row for that family containing fit results is added to the Comparison Details report.
- If Nonparametric is checked in the Distribution list, the Nonparametric Density report, showing the best fitting kernel bandwidth, is added to the Compare Distributions report. See "Nonparametric Density" on page 243.
- You can change your selected distribution by selecting its radio button under Selected. The capability report is updated to show results for the selected distribution.

Figure 10.16 shows the Compare Distributions report for Process 1 in the Process Measurements.jmp sample data table. The Selected distribution, which is Lognormal, is being compared to a Normal distribution. The Comparison Details report shows fit statistics for both distributions.

You can obtain probability plots by selecting the Probability Plots option from the Compare Distributions red triangle menu. The points in the probability plot for the normal distribution in Figure 10.16 do not follow the line closely. This indicates a poor fit.

Figure 10.16 Compare Distributions with Probability Plot for Normal

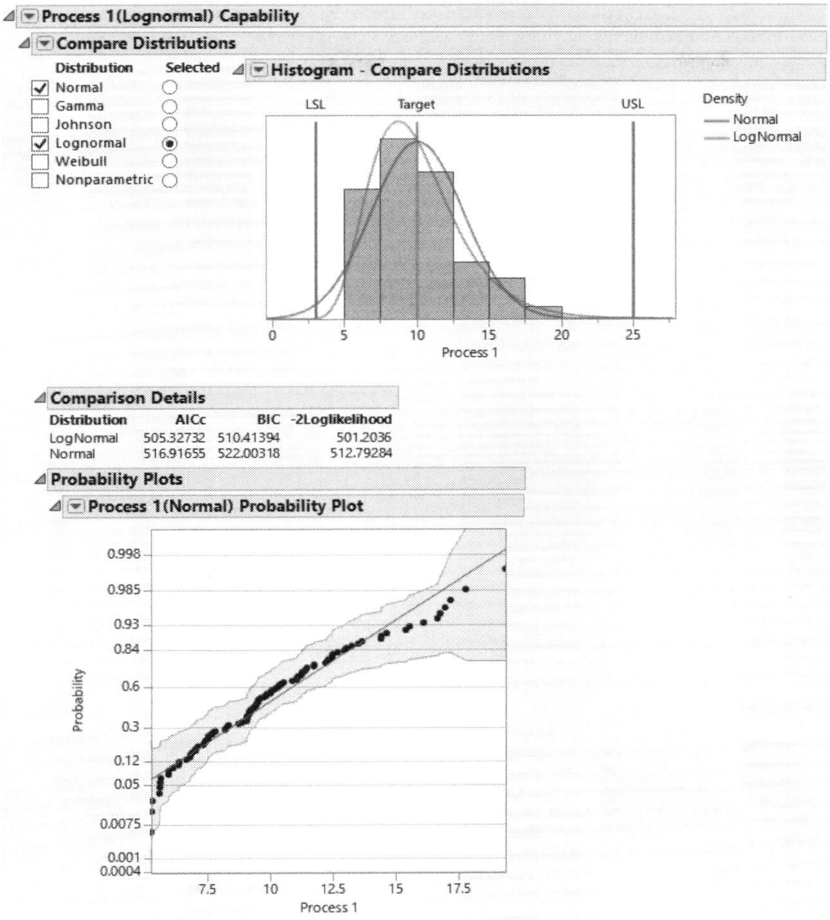

Distribution	AICc	BIC	-2Loglikelihood
LogNormal	505.32732	510.41394	501.2036
Normal	516.91655	522.00318	512.79284

Compare Distributions Options

The following options are available in the Compare Distributions red triangle menu.

Comparison Details For each distribution, gives AICc, BIC, and -2Loglikelihood values. See the Statistical Details appendix in the *Fitting Linear Models* book. (Not available for a Nonparametric fit.)

Comparison Histogram Shows or hides the Histogram report.

Probability Plots Shows or hides a report that displays probability plots for each parametric distribution that you fit. See Figure 10.16. An observation's horizontal coordinate is its mid-point adjusted Kaplan-Meier estimate. An observation's vertical coordinate is the value of the quantile of the fitted distribution for the observation's rank. For the normal distribution, the overall estimate of sigma is used in determining the fitted distribution.

Note: When it is not possible to calculate the quantiles for a probability scale for the gamma distribution, the plot shows a linear scale.

The red triangle menus associated with each Probability Plot contain the following options.

Simultaneous Empirical Confidence Limits Shows or hides confidence limits that have a simultaneous 95% confidence level of containing the true probability function, given that the data come from the selected parametric family. These limits have the same estimated precision at all points. Use them to determine whether the selected parametric distribution fits the data well. See Nair (1984) and Meeker and Escobar (1998).

Caution: The simultaneous empirical confidence limits are not affected by the selection of Alpha Level in the Process Capability launch window.

Simultaneous Empirical Confidence Limits Shading Shows or hides shading of the region between the Simultaneous Empirical Confidence Limits.

Parametric Fit Line Shows or hides the line that shows the predicted probabilities for the observations based on the fitted distribution.

Parametric Fit Confidence Limits Shading Shows or hides shading of the region between parametric fit confidence intervals. The parametric fit confidence limits have confidence level (1 - Alpha), where Alpha is the value that you specify in the launch window. (Available only when the parametric fit confidence limits are meaningful and when it is possible to calculate them.)

When possible, the intervals are computed by expressing the parametric distribution F as a location-scale family, so that $F(y) = G(z)$, where $z = (y - \mu)/\sigma$. The approximate standard error of the fitted location-scale component at a point is computed using the delta method. Using the standard error estimate, a Wald confidence interval for z is computed for each point. The confidence interval for the cumulative distribution function F is obtained by transforming the Wald interval using G. Note that, in some cases, special accommodations are required to provide appropriate intervals near the endpoints of the interval of process measurements.

Order by Comparison Criterion Orders the distributions in the Comparison Details report according to the criterion that you select. The default ordering is by AICc, unless you selected another criterion in the Distribution Comparison Criterion panel in the launch window.

Normalized Box Plots

The Within Sigma Normalized Box Plots and Overall Sigma Normalized Box Plots options show or hide box plots that have been normalized using the within sigma and overall sigma, respectively. When drawing normalized box plots, JMP standardizes each column by subtracting the mean and dividing by the standard deviation. The box plots are formed for each column using these standardized values.

Figure 10.17 Within Sigma Normalized Box Plot

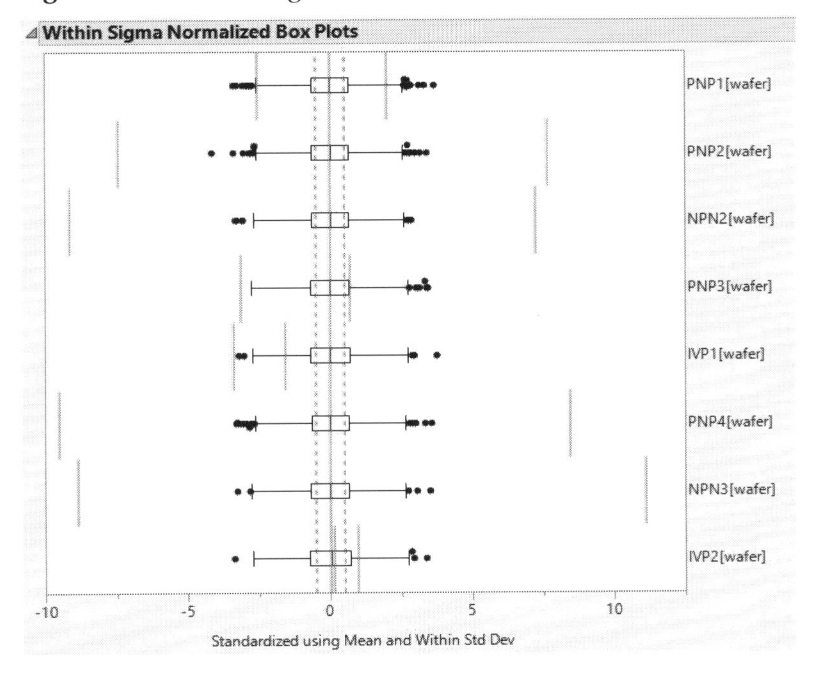

Figure 10.17 shows the Within Sigma Normalized Box Plot for a selection of the process variables in the Semiconductor Capability.jmp sample data table using wafer as a subgroup variable.

The green vertical lines represent the specification limits for each variable normalized by the mean and standard deviation of each variable. The gray dotted vertical lines are drawn at ±0.5, since the data is standardized to a standard deviation of 1.

Summary Reports

The Within Sigma Capability Summary Report and Overall Sigma Capability Summary Report options show or hide a table that contains the following statistics for each variable: LSL, Target, USL, Sample Mean, Sigma, Cpk, Cpl, Cpu, Cp, Cpm, and Nonconformance statistics. These statistics are calculated using the within sigma and overall sigma, respectively. Figure 10.18 shows a subset of columns for both summary reports as described in "Example of the Process Capability Platform with Normal Variables" on page 216. The following optional columns are available for this report:

- Confidence intervals for Cpk, Cpl, Cpu, CP, and Cpm
- Expected and observed PPM statistics (outside, below LSL, above USL)
- Sample standard deviation
- The sample size (N), the minimum, and the maximum.

To reveal these optional columns, right-click on the report and select the column names from the Columns submenu.

Note that the report (based on overall sigma) shows the overall capability indices Ppk, Ppl, Ppu, and Pp instead of the within capability indices Cpk, Cpl, Cpu, and Cp. The labeling of the overall capability indices depends on the setting of the AIAG (Ppk) Labeling preference.

Figure 10.18 Within Sigma and Overall Sigma Capability Summary Reports

Process Capability

Within Sigma Capability Summary Report

Process	LSL	Target	USL	Sample Mean	Within Sigma	Cpk	Cpl	Cpu	Cp	Cpm
PNP1	164.3895	297.0179	429.6463	313.0697	59.43013	0.654	0.834	0.654	0.744	0.718
PNP2	-136.122	465.442	1067.006	456.6157	79.27036	2.492	2.492	2.567	2.530	2.514
NPN2	96.59381	113.749	130.9042	115.7421	2.131652	2.371	2.994	2.371	2.683	1.959
PNP3	118.6778	130.2898	141.9018	137.6146	6.160912	0.232	1.025	0.232	0.628	0.404
IVP1	59.62007	63.41011	67.20015	73.78072	4.238298	-0.518	1.114	-0.518	0.298	0.113
PNP4	-54.4319	238.7386	531.9091	256.3756	33.22573	2.764	3.118	2.764	2.941	2.598
NPN3	97.31768	120.8047	144.2917	118.1352	2.362847	2.937	2.937	3.690	3.313	2.196
IVP2	139.2004	142.3052	145.4099	138.2432	7.406516	-0.043	-0.043	0.323	0.140	0.123

Overall Sigma Capability Summary Report

Process	LSL	Target	USL	Sample Mean	Overall Sigma	Ppk	Ppl	Ppu	Pp	Cpm
PNP1	164.3895	297.0179	429.6463	313.0697	58.35353	0.666	0.849	0.666	0.758	0.730
PNP2	-136.122	465.442	1067.006	456.6157	79.82589	2.475	2.475	2.549	2.512	2.497
NPN2	96.59381	113.749	130.9042	115.7421	2.100786	2.406	3.038	2.406	2.722	1.975
PNP3	118.6778	130.2898	141.9018	137.6146	6.060762	0.236	1.041	0.236	0.639	0.407
IVP1	59.62007	63.41011	67.20015	73.78072	4.196326	-0.523	1.125	-0.523	0.301	0.113
PNP4	-54.4319	238.7386	531.9091	256.3756	32.60738	2.817	3.177	2.817	2.997	2.636
NPN3	97.31768	120.8047	144.2917	118.1352	2.364757	2.934	2.934	3.687	3.311	2.195
IVP2	139.2004	142.3052	145.4099	138.2432	7.327164	-0.044	-0.044	0.326	0.141	0.124

Make Goal Plot Summary Table

The Make Goal Plot Summary Table option produces a summary data table that includes each variable's name, its spec-normalized mean shift (Mean Shift Standardized to Spec), and its spec-normalized standard deviation (Std Dev Standardized to Spec). For each variable, there is a row for each of the two sigma types (Within and Overall).

Note: If a variable is fit with a distribution other than normal, the name of the fitted distribution is appended parenthetically to the variable name. The Mean Shift Standardized to Spec and Std Dev Standardized to Spec values are not provided for nonnormal variables.

The points in the Goal Plot are linked to the rows in the Goal Plot Summary Table. If you apply row states to a point in the Goal Plot, you can change the corresponding row states in the Goal Plot Summary Table. Conversely, if you apply row states in the Goal Plot Summary Table, they are reflected on the Goal Plot.

Figure 10.19 shows the Goal Plot Summary Table for the Semiconductor Capability.jmp sample data table as described in "Example of the Process Capability Platform with Normal Variables" on page 216.

Figure 10.19 Summary Table

	Process	Sigma Type	Mean Shift Standardized to Spec	Std Dev Standardized to Spec
1	PNP1	Within	0.060514123	0.2240474981
2	PNP2	Within	-0.007336131	0.0658868743
3	NPN2	Within	0.0580913305	0.06212847
4	PNP3	Within	0.3153977755	0.2652827544
5	IVP1	Within	1.368139782	0.5591360695
6	PNP4	Within	0.0300797696	0.0566662245
7	NPN3	Within	-0.056828799	0.0503011804
8	IVP2	Within	-0.654156076	1.1927691323
9	PNP1	Overall	0.060514123	0.2199887772
10	PNP2	Overall	-0.007336131	0.0663486129
11	NPN2	Overall	0.0580913305	0.0612288463
12	PNP3	Overall	0.3153977755	0.2609703636
13	IVP1	Overall	1.368139782	0.5535989822
14	PNP4	Overall	0.0300797696	0.0556116319
15	NPN3	Overall	-0.056828799	0.0503418504
16	IVP2	Overall	-0.654156076	1.1799901131

Additional Examples of the Process Capability Platform

Process Capability for a Stable Process

In this example, you verify the assumptions that enable you to estimate PPM defective rates based on a capability analysis. You access Process Capability through Control Chart Builder and then directly. The data consist of 22 subgroups of size five. There are six missing readings, with three in each of two consecutive subgroups.

Process Capability through Control Chart Builder

You can use Control Chart Builder to check process stability and the normality assumption for your process characteristic. You can also obtain Process Capability information directly within Control Chart Builder.

1. Select **Help > Sample Data Library** and open Quality Control/Clips2.jmp.
2. Right-click the Gap column and select **Column Info**.
3. Select the **Spec Limits** column property.
4. Select **Show as graph reference lines** and click **OK**.
5. Select **Analyze > Quality and Process > Control Chart Builder**.

6. Drag Date to the **Subgroup** zone.
7. Drag Gap to the **Y** zone.

Figure 10.20 XBar and R Control Chart for Gap

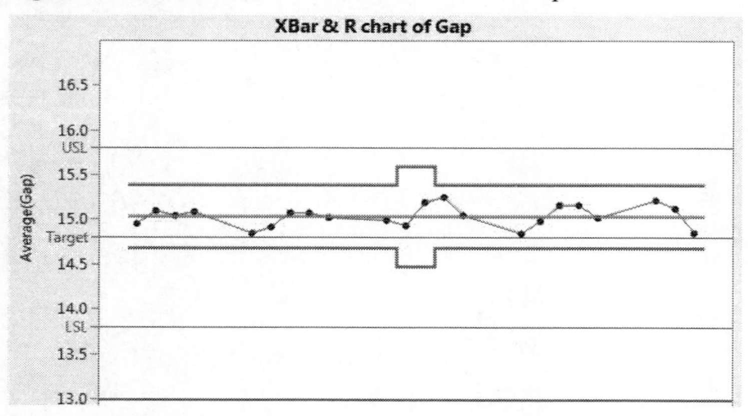

The control chart indicates that Gap is stable over time. Because Gap has the Spec Limits column property, a Process Capability Analysis report appears to the right of the control chart.

Figure 10.21 Histogram in Process Capability Analysis for Gap

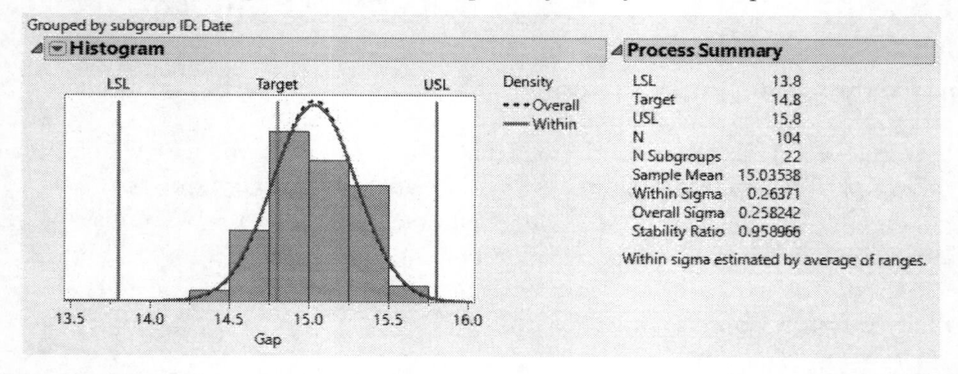

The histogram and fitted normal blue curve suggest that the distribution of Gap is approximately normal. Although the process is stable, the distribution of Gap is shifted to the right of the specification range.

The Process Summary report shows the specification limits that are saved to the Spec Limits column property. It also shows that the estimate of sigma calculated from within-subgroup variation (Within Sigma) does not differ greatly from the overall estimate given by the sample standard deviation (Overall Sigma). Consequently, the Stability Ratio is near one (0.958966). This is expected because the process is stable.

8. Right-click in the body of the Nonconformance report and select **Expected Within PPM** from the Columns submenu.

Figure 10.22 Capability Indices and Nonconformance Report

Within Sigma Capability

Index	Estimate	Lower 95%	Upper 95%
Cpk	0.966	0.805	1.128
Cpl	1.562	1.314	1.808
Cpu	0.966	0.805	1.127
Cp	1.264	1.071	1.457
Cpm	0.943	0.828	1.058

Overall Sigma Capability

Index	Estimate	Lower 95%	Upper 95%
Ppk	0.987	0.838	1.136
Ppl	1.595	1.367	1.821
Ppu	0.987	0.837	1.135
Pp	1.291	1.115	1.467
Cpm	0.954	0.841	1.072

Nonconformance

Portion	Observed %	Expected Within %	Expected Overall %	Expected Within PPM
Below LSL	0.0000	0.0001	0.0001	1.402263
Above USL	0.0000	0.1869	0.1534	1869.0329
Total Outside	0.0000	0.1870	0.1535	1870.4352

The Cpk value calculated using subgroup variation is 0.966, indicating that the process is not very capable. The Cpl value suggests good performance, but this is because the process is shifted away from the lower specification limit. Defective parts generally result from large values of Gap.

Note that the confidence interval for Cpk is wide; it ranges from 0.805 to 1.128. This occurs even though there are 104 observations. Capability indices are surprisingly variable, due to the fact that they are ratios. It is easy to reach incorrect conclusions based on the point estimate of a capability index.

The estimates of out-of-specification product given in the Nonconformance report provide a direct measure of process performance. The PPM values in the Nonconformance report indicate that Gap hardly ever falls below the lower specification limit (1.4 parts per million). However, the number of parts for which Gap falls above the upper specification limit is 1869.0 parts per million.

For an uncentered process, the Cp value indicates potential capability if the process were adjusted to be centered. If this process were adjusted to be centered at the target value of 14.8, then its capability would be 1.264, with a confidence interval from 1.071 to 1.457.

Process Capability Platform

Now that you have verified stability and normality for Gap, you can obtain additional information in the Process Capability platform.

1. Select **Analyze > Quality and Process > Process Capability**.
2. Select Gap and click **Y, Process**.
3. Open the **Process Subgrouping** outline.
4. Select Date in the Select Columns list and Gap in the Roles list.
5. Click **Nest Subgroup ID Column**.

By default, the Within-Subgroup Variation Statistic selection is set to Average of Unbiased Standard Deviations. In the Control Chart Builder example ("Process Capability through Control Chart Builder" on page 249), subgroup ranges were used.

6. Click **OK.**

Figure 10.23 Goal Plot and Box Plot for Gap

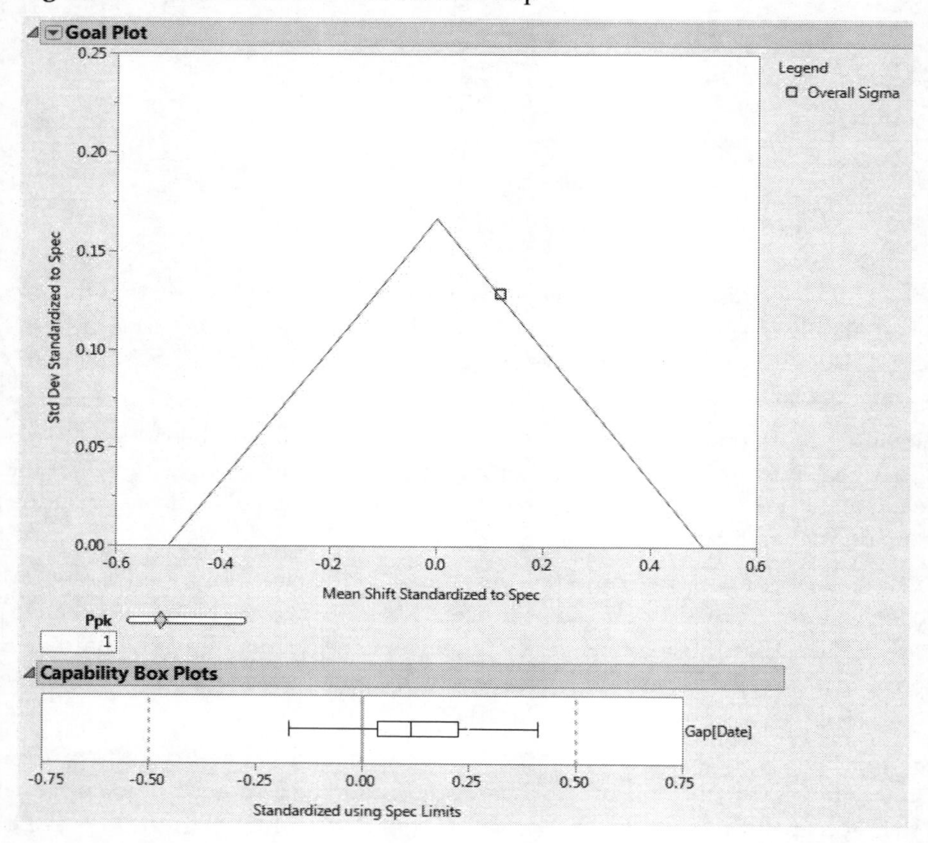

The Goal Plot shows the Ppk index for Gap as being essentially equal to 1. The box plot shows that most values fall within specifications, but the preponderance of data values are shifted to the right within the specification range.

7. From the Process Capability red triangle menu, select Individual Detail Reports.

   The report is the one obtained using Control Chart Builder, except that the Within Sigma is based on average standard deviations rather than average ranges. See "Histogram in Process Capability Analysis for Gap" on page 250 and "Capability Indices and Nonconformance Report" on page 251.

Process Capability for an Unstable Process

The following example shows a case where the overall variation differs from the within variation because the process is not stable. It uses the **Coating.jmp** data table from the **Quality Control** folder of Sample Data (taken from the *ASTM Manual on Presentation of Data and Control Chart Analysis*). The process variable of interest is the **Weight** column, grouped into subgroups by the **Sample** column.

Process Capability Platform

1. Select **Help > Sample Data Library** and open Quality Control/Coating.jmp.
2. Select **Analyze > Quality and Process > Process Capability**.
3. Select Weight and click **Y, Process**.
4. Open the **Process Subgrouping** outline.
5. Select Sample in the **Select Columns** list on the left.
6. Select Weight in the **Cast Selected Columns into Roles** list on the right.
7. Click **Nest Subgroup ID Column**.
8. Click **OK**.
9. Enter 16 for **LSL**, 20 for **Target**, and 24 for **USL** in the **Spec Limits** window.
10. Click **OK**.
11. Select **Show Within Sigma Points** from the Goal Plot red triangle menu.
12. Select **Individual Detail Reports** from the Process Capability red triangle menu.

Figure 10.24 Process Capability Report for Coating.jmp Data

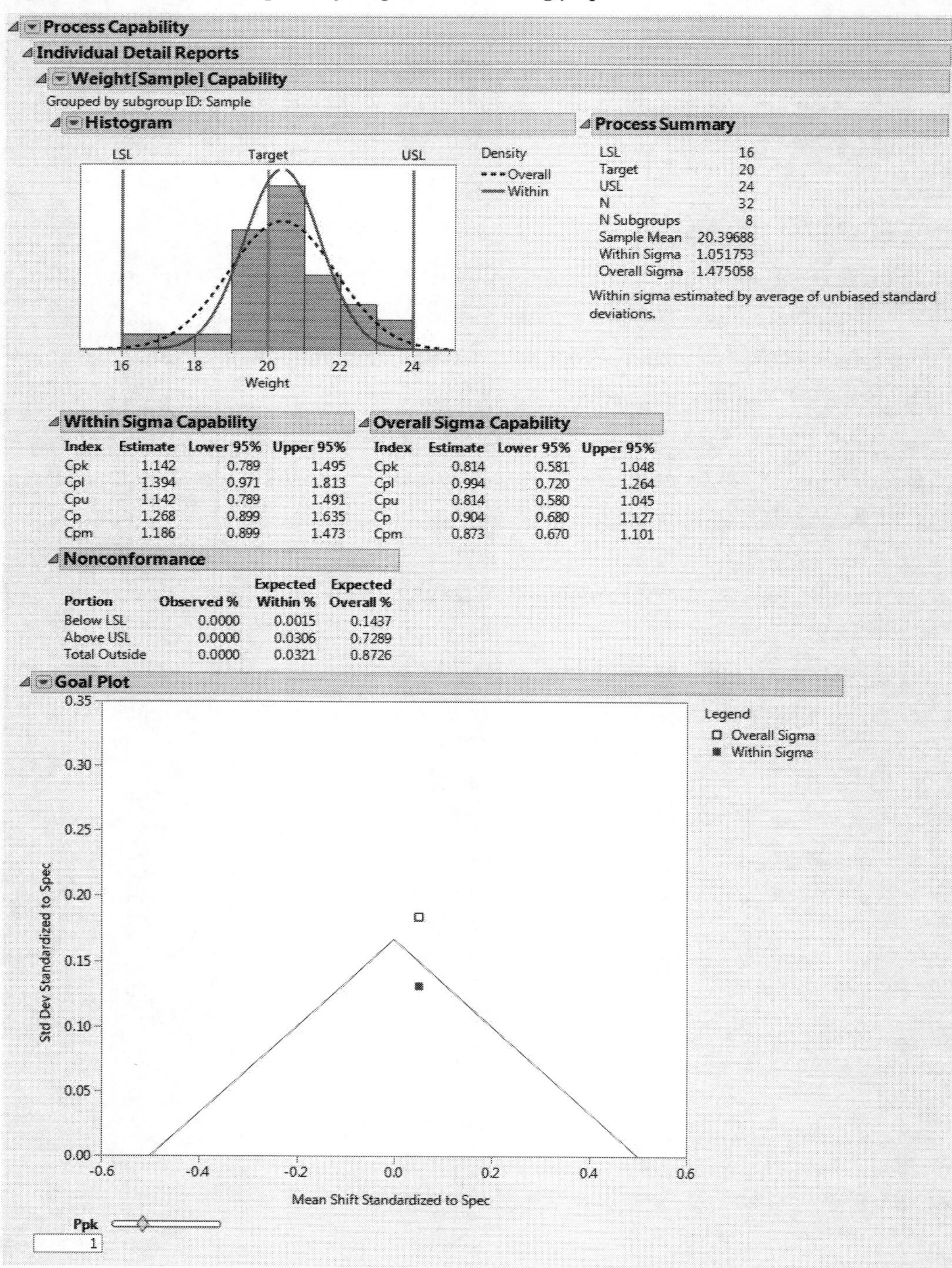

Figure 10.24 shows the resulting Process Capability report. The Goal Plot shows two points that represent the mean shift and standard deviation standardized to the specification limits. The point labeled Overall Sigma is calculated using the overall sample standard deviation. The point labeled Within Sigma is calculated using a within-subgroup estimate of the standard deviation.

The point calculated using Overall Sigma is outside the goal triangle corresponding to a Ppk of 1. This indicates that the variable Weight results in non-conforming product.

However, the point calculated using Within Sigma is inside the goal triangle. This indicates that, if the process were stable, Weight values would have a high probability of falling within the specification limits.

Control Chart to Assess Stability

Use Control Chart Builder to determine whether the Weight measurements are stable.

1. Select **Help > Sample Data Library** and open Quality Control/Coating.jmp.
1. Select **Analyze > Quality and Process > Control Chart Builder**.
2. Drag Sample to the **Subgroup** zone.
3. Drag Weight to the **Y** zone.

Figure 10.25 XBar and R Chart for Weight

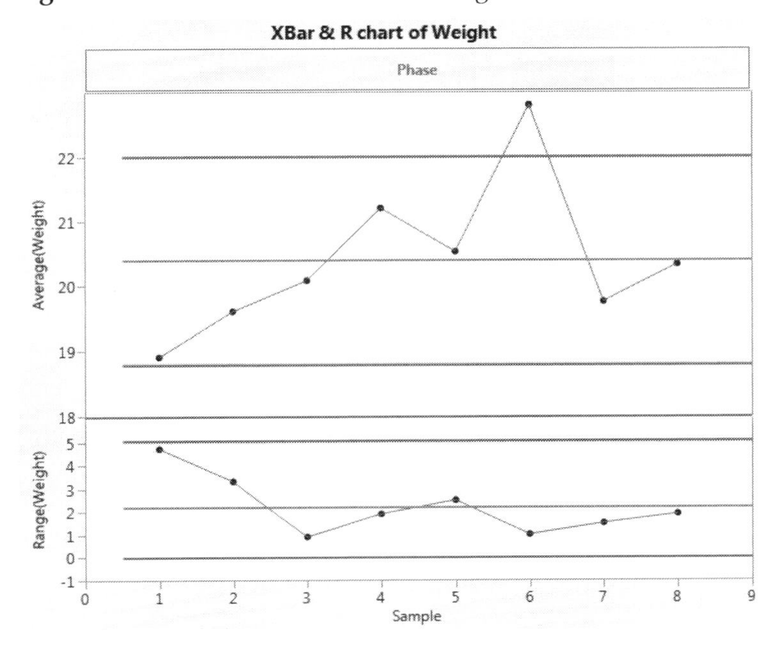

The control chart indicates that the Weight measurements are unstable. The process is affected by special causes and is unpredictable. This makes the interpretation of capability indices and nonconformance estimates highly questionable. Even estimates based on Overall Sigma are questionable, because the process is not predictable.

The histogram in Figure 10.24 shows the distribution of the Weight values with normal density curves using both sigma estimates superimposed over the histogram. The normal curve that uses the Overall Sigma estimate is flatter and wider than the normal curve that uses the Within Sigma estimate. This normal curve is more dispersed because the estimate of Overall Sigma is inflated by the special causes that make the process unstable. If the process were stable, the narrower normal curve would reflect process behavior.

You can also compare the Cpk estimate (1.142) to the Ppk estimate (0.814). The fact that Ppk is much smaller than Cpk is additional evidence that this is an unpredictable process. The Cpk estimate is a forecast of the capability that you would achieve by bringing the process to a stable state.

Note: The Individual Detail Reports Cutoff preference determines whether the Individual Reports appear by default. If the preference is enabled, the Individual Reports appear by default if the number of process variables is less than or equal to the number specified in the preference. You can change this preference in Preferences > Platforms > Process Capability.

Simulation of Confidence Limits for a Nonnormal Process Ppk

In this example, you first perform a capability analysis for the three nonnormal variables in Tablet Measurements.jmp. You then use Simulate to find confidence limits for the nonconformance percentage for the variable Purity.

Nonnormal Capability Analysis

If you prefer not to follow the steps below, you can obtain the results in this section by running the **Process Capability** table script in Tablet Measurements.jmp.

1. Select **Help > Sample Data Library** and open Tablet Measurements.jmp.
2. Select **Analyze > Quality and Process > Process Capability.**
3. Select Weight, Thickness, and Purity and click **Y, Process.**
4. Select Weight, Thickness, and Purity in the **Cast Selected Columns into Roles** list on the right.
5. Open the **Distribution Options** outline.
6. From the Distribution list, select **Best Fit.**
7. Click **Set Process Distribution.**

   The **&Dist(Best Fit)** suffix is added to each column name in the list on the right.

8. Click **OK**.

   A Capability Index Plot appears, showing the Ppk values. Note that only the Thickness variable appears above the line that denotes Ppk = 1. Purity is nearly on the line. Although the number of measurements, 250, seems large, the estimated Ppk value is still quite variable. For this reason, you construct a confidence interval for the true Purity Ppk value.

   Note: Because a Goal Plot is not shown, you can conclude that a normal distribution fit was not the best fit for any of the three variables.

9. Select **Individual Detail Reports** from the Process Capability red triangle menu.

   The best fits are as follows:

   – Weight: Lognormal
   – Thickness: Johnson Sb (see the note immediately beneath the Thickness(Johnson) Capability report title)
   – Purity: Weibull

Construct the Simulation Column

To use the Simulate utility to estimate Ppk confidence limits, you need to construct a simulation formula that reflects the fitted Weibull distribution. If you prefer not to follow the steps below, you can obtain the results in this section by running the **Add Simulation Column** table script.

1. Scroll to the Purity (Weibull) Capability report and find the Parameter Estimates report.

Figure 10.26 Weibull Parameter Estimates for Purity

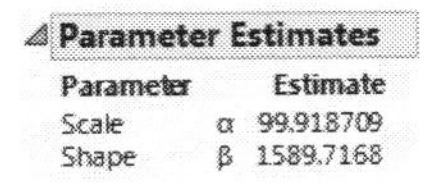

Parameter Estimates

Parameter		Estimate
Scale	α	99.918709
Shape	β	1589.7168

These are the parameter estimates for the best fitting distribution, which is Weibull.

1. In the Tablet Measurements.jmp sample data table, select **Cols > New Columns**.
2. Next to **Column Name**, enter Simulated Purity.
3. From the **Column Properties** list, select **Formula**.
4. In the formula editor, select **Random > Random Weibull**.
5. The placeholder for **beta** is selected. Click the insertion element (^).

Figure 10.27

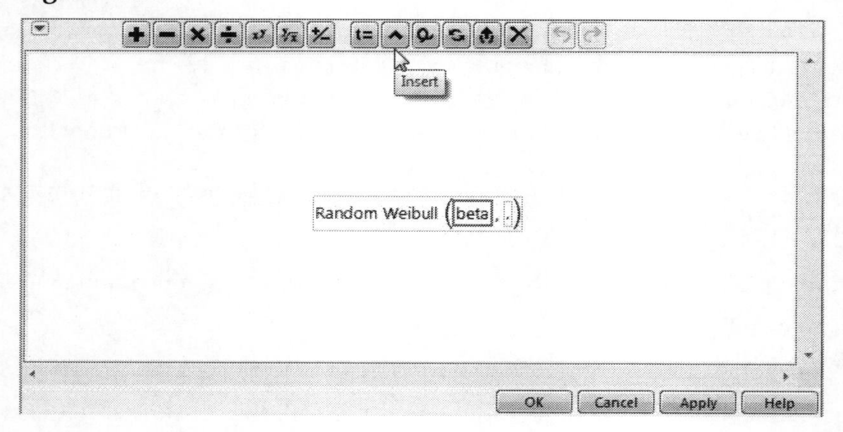

This adds a placeholder for the parameter **alpha**.

6. Right-click in the Parameter Estimates report table and select **Make into Data Table**.
7. Copy the entry in Row 2 in the Estimate column (1589.7167836).
8. In the formula editor window, select the placeholder for **beta** in the **Random Weibull** formula and paste 1589.7167836 into the placeholder for **beta**.
9. In the data table that you created from the Parameter Estimates report, copy the entry in Row 1 in the Estimate column (99.918708989).
10. In the formula editor window, select the placeholder for **alpha** in the **Random Weibull** formula and paste 99.918708989 into the placeholder for **alpha**.

Figure 10.28 Completed Formula Window

Random Weibull (1589.7167836, 99.918708989)

11. Click **OK** in the formula editor window.

    The Simulated Purity column contains a formula that simulates values from the best-fitting distribution.

Simulate Confidence Intervals for Purity Ppk and Expected % Nonconforming

When you use Simulate, the entire analysis is run the number of times that you specify. To shorten the computing time, you can minimize the computational burden by running only the required analysis. In this example, because you are interested only in Purity with a fitted Weibull distribution, you perform only this analysis before running Simulate.

1. In the Process Capability report, select **Relaunch Dialog** from the Process Capability red triangle menu.

2. (Optional) Close the Process Capability report.
3. In the launch window, from the **Cast Selected Columns into Roles** list, select Weight&Dist(Lognormal) and Thickness&Dist(Johnson).
4. Click **Remove**.
5. Click **OK**.
6. Select **Individual Detail Reports** from the Process Capability red triangle menu,

   Both Ppk and Ppl values are provided, but they are identical because Purity has only a lower specification limit.
7. In the Overall Sigma Capability report, right-click the **Estimate** column and select **Simulate**.

   In the **Column to Switch Out** list, Purity is selected. In the **Column to Switch In** list, Simulated Purity is selected.
8. Next to **Number of Samples**, enter **500**.

Note: The next step is not required. However, it ensures that you obtain exactly the simulated values shown in this example.

---

9. (Optional) Next to **Random Seed**, enter **12345**.
10. Click **OK**.

    The calculation might take several seconds. A data table entitled Process Capability Simulate Results (Estimate) appears. The Ppk and Ppl columns in this table each contain 500 values calculated based on the Simulated Purity formula. The first row, which is excluded, contains the values for Purity obtained in your original analysis. Because Purity has only a lower specification limit, the Ppk values are identical to the Ppl values.
11. In the Process Capability Simulate Results (Estimate) data table, click the green triangle next to the **Distribution** script.

Figure 10.29 Distribution of Simulated Ppk Values for Purity

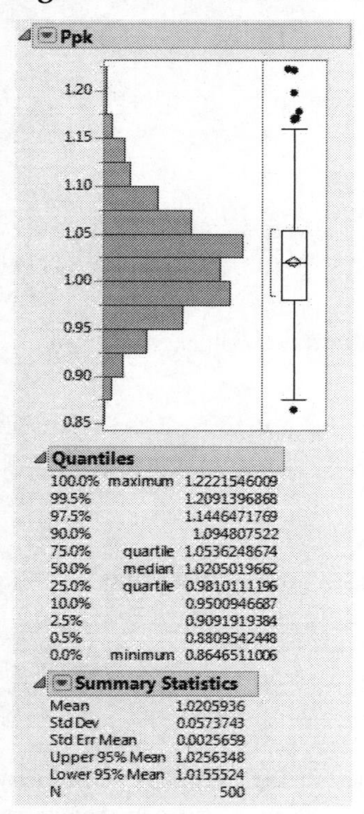

Two Distribution reports are shown, one for Ppk and one for Ppl. But **Purity** has only a lower specification limit, so that the Ppk and Ppl values are identical. For this reason, the Distribution reports are identical.

A 95% confidence interval for Ppk based on the 2.5% and 97.5% quantiles is 0.9092 to 1.1446. The true Ppk value could be above 1.0, which would place **Purity** above the Ppk = 1 line in the Capability Index Plot you constructed in "Nonnormal Capability Analysis" on page 256.

12. In the Process Capability report, right-click the **Expected Overall %** column in the Nonconformance report and select **Simulate**.
13. Next to **Number of Samples**, enter **500**.
14. (Optional) Next to **Random Seed**, enter **12345**.
15. Click **OK**.

The calculation might take several seconds. A data table entitled Process Capability Simulate Results (Expected Overall %) appears. Because **Purity** has only a lower specification limit, the **Below LSL** values are identical to the **Total Outside** values.

16. In the Process Capability Simulate Results (Expected Overall %) data table, click the green triangle next to the **Distribution** script.

Figure 10.30 Distribution of Simulated Total Outside Values for Purity

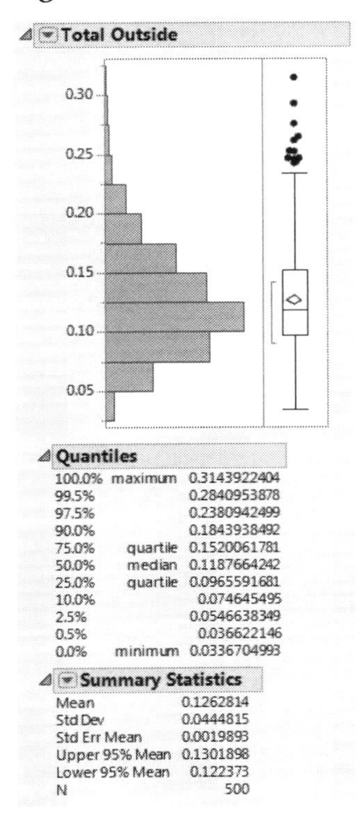

Again, two identical Distribution reports appear. A 95% confidence interval for the Expected Overall % nonconforming, based on the 2.5% and 97.5% quantiles, is 0.055 to 0.238.

Statistical Details for the Process Capability Platform

The following sections provide statistical details for the Process Capability platform:

- "Variation Statistics" on page 262
- "Notation for Goal Plots and Capability Box Plots" on page 264
- "Goal Plot" on page 264
- "Capability Box Plots for Processes with Missing Targets" on page 266
- "Capability Indices for Normal Distributions" on page 267

- "Capability Indices for Nonnormal Distributions: Percentile and Z-Score Methods" on page 267
- "Parameterizations for Distributions" on page 269

Variation Statistics

Denote the standard deviation of the process by σ. The Process Capability platform provides two types of capability indices. The Ppk indices are based on an estimate of σ that uses all of the data in a way that does not depend on subgroups. This overall estimate can reflect special cause as well as common cause variation. The Cpk indices are based on an estimate that attempts to capture only common cause variation. The Cpk indices are constructed using within-subgroup, or short-term, estimates of σ. In this way, they attempt to reflect the true process standard deviation. When a process is not stable, the overall and within estimates of σ can differ markedly.

Overall Sigma

The overall sigma does not depend on subgroups. JMP calculates the overall estimate of σ as follows:

$$\hat{\sigma} = \frac{1}{N-1}\sum_{i=1}^{N}(y_i - \bar{y})^2$$

The formula uses the following notation:

N = number of nonmissing values in the entire data set

y_i = value of the ith observation

$\bar{y}$ = mean of nonmissing values in the entire data set

Caution: When the process is stable, the Overall Sigma estimates the process standard deviation. If the process is not stable, the overall estimate of σ is of questionable value, since the process standard deviation is unknown.

Estimates of Sigma Based on Within-Subgroup Variation

An estimate of σ that is based on within-subgroup variation can be constructed in one of three ways:

- Within sigma estimated by average of ranges
- Within sigma estimated by average of unbiased standard deviations
- Within sigma estimated by moving range

If you specify a subgroup ID column or a constant subgroup size on the launch window, you can specify your preferred within-subgroup variation statistic. See "Launch the Process Capability Platform" on page 223. If you do not specify a subgroup ID column, a constant subgroup size, or a historical sigma, JMP estimates the within sigma using the third method (moving range of subgroups of size two).

Within Sigma Based on Average of Ranges

Within sigma estimated by the *average of ranges* is the same as the estimate of the standard deviation of an $\overline{X}/R$ chart:

$$\hat{\sigma} = \frac{\frac{R_1}{d_2(n_1)} + \ldots + \frac{R_N}{d_2(n_N)}}{N}$$

The formula uses the following notation:

R_i = range of ith subgroup

n_i = sample size of ith subgroup

$d_2(n_i)$ = expected value of the range of n_i independent normally distributed variables with unit standard deviation

N = number of subgroups for which $n_i \geq 2$

Within Sigma Based on Average of Unbiased Standard Deviations

Within sigma estimated by the *average of unbiased standard deviations* is the same as the estimate for the standard deviation in an $\overline{X}/S$ chart:

$$\hat{\sigma} = \frac{\frac{s_1}{c_4(n_1)} + \ldots + \frac{s_N}{c_4(n_N)}}{N}$$

The formula uses the following notation:

n_i = sample size of ith subgroup

$c_4(n_i)$ = expected value of the standard deviation of n_i independent normally distributed variables with unit standard deviation

N = number of subgroups for which $n_i \geq 2$

s_i = sample standard deviation of the ith subgroup.

Within Sigma Based on Moving Range

Within sigma estimated by *moving range* is the same as the estimate for the standard deviation for Individual Measurement and Moving Range charts:

$$\hat{\sigma} = \frac{\overline{MR}}{d_2(2)}$$

The formula uses the following notation:

$\overline{MR}$ = the mean of the nonmissing moving ranges computed as $(MR_2+MR_3+...+MR_N)/(N-1)$ where $MR_i = |y_i - y_{i-1}|$.

$d_2(2)$ = expected value of the range of two independent normally distributed variables with unit standard deviation.

Notation for Goal Plots and Capability Box Plots

The formulas for the Goal Plot and Capability Box Plots use the following notation:

Y_{ij} = i^{th} observation for process j

$\bar{Y}_j$ = mean of the observations on process j

$SD(Y_j)$ = standard deviation of the observations on process j

T_j = target value for process j

LSL_j = lower specification limit for process j

USL_j = upper specification limit for process j

$SD(Y_j)$ = standard deviation for process j

Goal Plot

This section provides details about the calculation of the mean shift and standard deviation standardized to specification quantities plotted in the Goal Plot. This section uses the notation defined in "Notation for Goal Plots and Capability Box Plots" on page 264.

The mean shift and the standard deviation standardized to the specification limits for the jth column are defined as follows:

$$\text{Mean Shift Standardized to Spec} = \frac{\bar{Y}_j - T_j}{2 \times \min(T_j - LSL_j, USL_j - T_j)}$$

$$\text{Std Dev Standardized to Spec} = \frac{SD(Y_j)}{2 \times \min(T_j - LSL_j, USL_j - T_j)}$$

Note: If either LSL_j or USL_j is missing, twice the distance from the target to the nonmissing specification limit is used in the denominators of the Goal Plot coordinates.

Goal Plot Points for Processes with Missing Targets

Suppose that the process has both a lower and an upper specification limit but no target. Then the formulas given in "Goal Plot" on page 264 are used, replacing T_j with the midpoint of the two specification limits.

Suppose that the process has only one specification limit and no target. To obtain (x,y) coordinates for a point on the Goal Plot, the capability indices of the process are used. (See "Capability Indices for Normal Distributions" on page 267 for definitions in terms of the theoretical mean and standard deviation.) For sample observations, the following relationships hold:

$$C_{pu} = \frac{USL_j - \bar{Y}_j}{3SD(Y_j)}$$

$$C_{pl} = \frac{\bar{Y}_j - LSL_j}{3SD(Y_j)}$$

If a process has two specification limits and a target at the midpoint of the limits, then the (x,y) coordinates for the point on the Goal Plot satisfy these relationships:

$$C_{pu} = (0.5 - x)/3y$$

$$C_{pl} = (0.5 + x)/3y$$

To obtain coordinates when there is only one specification limit and no target, these relationships are used. To identify a unique point requires an assumption about the slope of the line from the origin on which the points fall. A slope of 0.5 is assumed for the case of an upper specification limit and of -0.5 for a lower specification limit. When capability values are equal to one and the Ppk slider for the goal plot triangle is set to 1, these slopes place the points on the goal plot triangle lines.

Consider the case of only an upper specification limit and no target. Using the assumption that the (x,y) coordinates fall on a line from the origin with slope 0.5, solving for x and y gives the following coordinates:

$$x = 1/(3C_{pu} + 2)$$
$$y = 1/(6C_{pu} + 4)$$

Consider the case of only a lower specification limit and no target. Using the assumption that the (x,y) coordinates fall on a line from the origin with slope -0.5, solving for x and y gives the following coordinates:

$$x = -1/(3C_{pl} + 2)$$
$$y = 1/(6C_{pl} + 4)$$

Note: If either C_{pu} or C_{pl} is less than -0.6, then it is set to -0.6 in the formulas above. At the value -2/3, the denominator for x assumes the value 0. Bounding the capability values at -0.6 prevents the denominator from assuming the value 0 or switching signs.

Capability Box Plots for Processes with Missing Targets

A column with no target can have both upper and lower specification limits, or only a single specification limit. This section uses the notation defined in "Notation for Goal Plots and Capability Box Plots" on page 264.

Two Specification Limits and No Target

When no target is specified for the jth column, the capability box plot is based on the following values for the transformed observations:

$$Z_{ij} = \frac{Y_{ij} - (\mathrm{LSL}_j + \mathrm{USL}_j)/2}{\mathrm{USL}_j - \mathrm{LSL}_j}$$

Single Specification Limit and No Target

Suppose that only the lower specification limit is specified. (The case where only the upper specification limit is specified in a similar way.)

When no target is specified for the jth column, the capability box plot is based on the following values for the transformed observations:

$$Z_{ij} = \frac{Y_{ij} - \bar{Y}_j}{2(\bar{Y}_j - \mathrm{LSL}_j)}$$

Note: When a column has only one specification limit and no target value, and the sample mean falls outside the specification interval, no capability box plot for that column is plotted.

Capability Indices for Normal Distributions

This section provides details about the calculation of capability indices for normal data.

For a process characteristic with mean μ and standard deviation σ, the population-based capability indices are defined as follows. For sample observations, the parameters are replaced by their estimates:

$$\text{Cp} = \frac{USL - LSL}{6\sigma}$$

$$\text{Cpl} = \frac{\mu - LSL}{3\sigma}$$

$$\text{Cpu} = \frac{USL - \mu}{3\sigma}$$

$$\text{Cpk} = \min(\text{Cpl}, \text{Cpu})$$

$$\text{Cpm} = \frac{USL - LSL}{6\sigma\sqrt{1 + \left(\frac{T - \mu}{\sigma}\right)^2}}$$

The formulas use the following notation:

LSL = Lower specification limit

USL = Upper specification limit

T = Target value

For estimates of Within Sigma capability, σ is estimated using the subgrouping method that you specified. For estimates of Overall Sigma capability, σ is estimated using the sample standard deviation. With the default AIAG (Ppk) Labeling, the indices based on Overall Sigma are denoted by Pp, Ppl, Ppu, and Ppk. The labeling for the index Cpm does not change when Overall Sigma is used.

If either of the specification limits is missing, the capability indices containing the missing specification limit are reported as missing.

Capability Indices for Nonnormal Distributions: Percentile and Z-Score Methods

This section describes how capability indices are calculated for nonnormal distributions. Two methods are described: the Percentile (also known as ISO/Quantile) method and the Z-Score (also known as Bothe/Z-scores) method. When you select a distribution for a nonnormal process variable, you can fit a parametric distribution or a nonparametric distribution. You can use either the Percentile or the Z-Score methods to calculate capability indices for the process variable of interest. However, unless you have a very large amount of data, a nonparametric fit might not accurately reflect behavior in the tails of the distribution.

Note: For both the Percentile and the Z-Score methods, if the data are normally distributed, the capability formulas reduce to the formulas for normality-based capability indices.

The descriptions of the two methods use the following notation:

LSL = Lower specification limit

USL = Upper specification limit

T = Target value

Percentile (ISO/Quantile) Method

The percentile method replaces the mean in the standard capability formulas with the median of the fitted distribution and the 6σ range of values with the corresponding percentile range. The method is described in AIAG (2005).

Denote the α*100th percentile of the fitted distribution by P_α. Then Percentile method capability indices are defined as follows:

$$P_{pk} = \min\left(\frac{P_{0.5} - LSL}{P_{0.5} - P_{0.00135}}, \frac{USL - P_{0.5}}{P_{0.99865} - P_{0.5}}\right)$$

$$P_{pl} = \frac{P_{0.5} - LSL}{P_{0.5} - P_{0.00135}}$$

$$P_{pu} = \frac{USL - P_{0.5}}{P_{0.99865} - P_{0.5}}$$

$$P_p = \frac{USL - LSL}{P_{0.99865} - P_{0.00135}}$$

$$C_{pm} = \frac{\min\left(\frac{T - LSL}{P_{0.5} - P_{0.00135}}, \frac{USL - T}{P_{0.99865} - P_{0.5}}\right)}{\sqrt{1 + \left(\frac{\mu - T}{\sigma}\right)^2}}$$

Z-Score (Bothe/Z-Scores) Method

The Z-Score method transforms the specification limits to values that have the same probabilities on a standard normal scale. It computes capability measures that correspond to a normal distribution with the same risk levels as the fitted nonnormal distribution.

Let F denote the fitted distribution for a process variable with lower and upper specification limits given by LSL and USL. Define equivalent standard normal specification limits as follows:

$$\begin{bmatrix} LSL_F = \Phi^{-1}(F(LSL)) \\ USL_F = \Phi^{-1}(F(USL)) \end{bmatrix}$$

Then the Z-Score method capability indices are defined as follows:

$$P_{pk} = \min(-LSL_F/3, USL_F/3)$$

$$P_{pl} = -LSL_F/3$$

$$P_{pu} = USL_F/3$$

$$P_p = (USL_F - LSL_F)/6$$

Note: Because Cpm is a target-based measure, it cannot be calculated using the Z-Scores method.

Note: For very capable data, *F(LSL)* or *F(USL)* can be so close to zero or one, respectively, that LSL_F or USL_F cannot be computed. In these cases, the corresponding capability indices are defined as Infinity.

Parameterizations for Distributions

This section gives the density functions *f* for the distributions used in the Process Capability platform. It also gives expected values and variances for all but the Johnson distributions.

Normal

$$f(x|\mu, \sigma) = \frac{1}{\sigma\sqrt{2\pi}}\exp\left[-\frac{1}{2\sigma^2}(x-\mu)^2\right], -\infty < x < \infty, -\infty < \mu < \infty, \sigma > 0$$

$$E(X) = \mu$$

$\text{Var}(X) = \sigma^2$

Gamma

$$f(x|\alpha, \sigma) = \frac{1}{\Gamma(\alpha)\sigma^{\alpha}} x^{\alpha - 1} \exp(-x/\sigma), x > 0, \alpha > 0, \sigma > 0$$

$\text{E}(X) = \alpha\sigma$

$\text{Var}(X) = \alpha\sigma^2$

Johnson

Johnson Su

$$f(x|\gamma, \delta, \sigma, \theta) = \frac{\delta}{\sigma}\left[1 + \left(\frac{x-\theta}{\sigma}\right)^2\right]^{-1/2} \phi\left[\gamma + \delta \sinh^{-1}\left(\frac{x-\theta}{\sigma}\right)\right], -\infty < x, \theta, \gamma < \infty, \theta > 0, \delta > 0$$

Johnson Sb

$$f(x|\gamma, \delta, \sigma, \theta) = \phi\left[\gamma + \delta \ln\left(\frac{x-\theta}{\sigma-(x-\theta)}\right)\right]\left(\frac{\delta\sigma}{(x-\theta)(\sigma-(x-\theta))}\right), \theta < x < \theta{+}\sigma, \sigma > 0$$

Johnson Sl

$$f(x|\gamma, \delta, \sigma, \theta) = \frac{\delta}{|x-\theta|}\phi\left[\gamma + \delta \ln\left(\frac{x-\theta}{\sigma}\right)\right], \text{ for } \; x > \theta \text{ if } \sigma = 1, x < \theta \text{ if } \sigma = -1$$

where $\phi(\cdot)$ is the standard normal probability density function.

Lognormal

$$f(x|\mu, \sigma) = \frac{1}{\sigma\sqrt{2\pi}} \frac{\exp\left[\frac{-(\log(x)-\mu)^2}{2\sigma^2}\right]}{x}, x > 0, -\infty < \mu < \infty, \sigma > 0$$

$\text{E}(X) = \exp(\mu + \sigma^2/2)$

$\text{Var}(X) = \exp(2(\mu + \sigma^2)) - \exp(2\mu + \sigma^2)$

Weibull

$$f(x|\alpha, \beta) = \frac{\beta}{\alpha^{\beta}} x^{\beta - 1} \exp\left[-\left(\frac{x}{\alpha}\right)^{\beta}\right], \alpha > 0, \beta > 0$$

$$E(X) = \alpha\Gamma\left(1 + \frac{1}{\beta}\right)$$

$$Var(X) = \alpha^2\left\{\Gamma\left(1 + \frac{2}{\beta}\right) - \Gamma^2\left(1 + \frac{1}{\beta}\right)\right\}$$, where $\Gamma(\cdot)$ is the gamma function.

Chapter 11

Capability Analysis

Evaluate the Ability of a Process to Meet Specifications

Capability analysis, used in process control, measures the conformance of a process to given specification limits. Using these limits, you can compare a current process to specific tolerances and maintain consistency in production. Graphical tools such as a goal plot and box plot give you quick visual ways of observing behaviors that are within specifications. Individual detail reports display a capability report for each variable in the analysis. The analysis helps reduce the variation relative to the specifications or requirements, achieving increasingly higher conformance values.

Figure 11.1 Examples of the Capability Platform

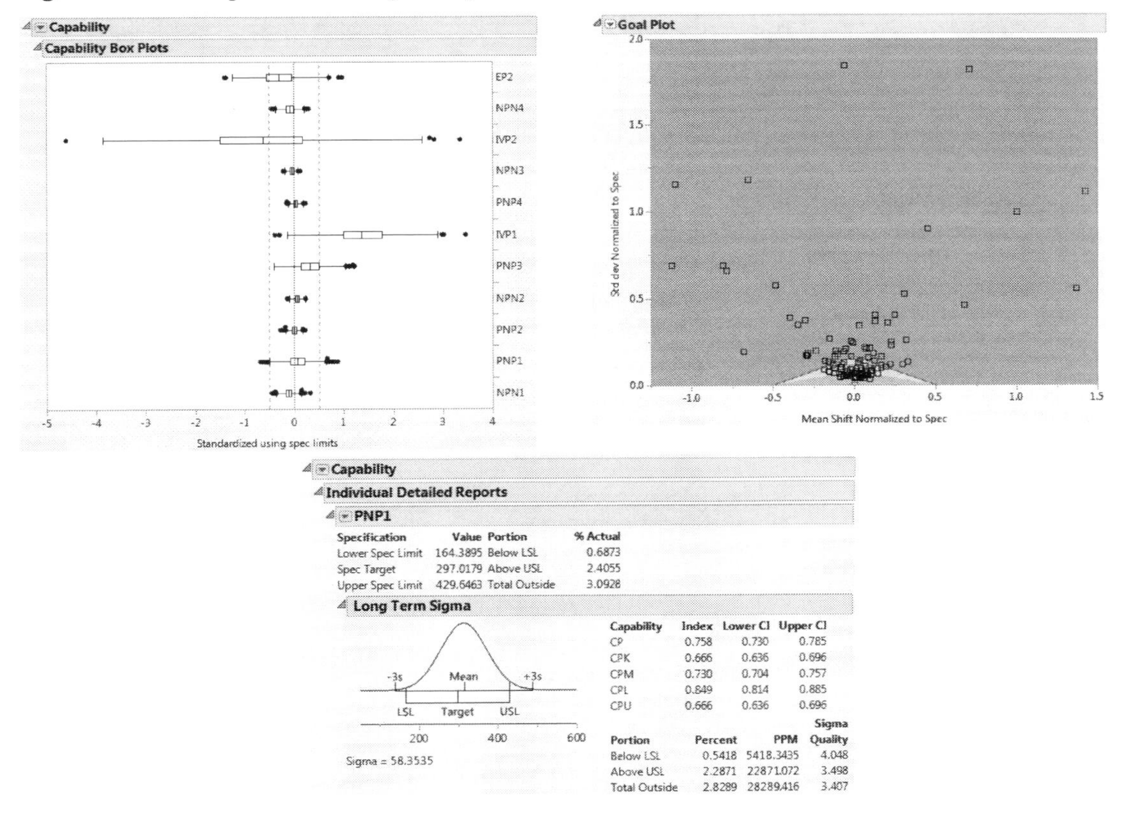

Overview of the Capability Platform

The Capability platform provides the tools needed to measure the compliance of a process to given specifications. By default, JMP shows a goal plot and capability box plot for each variable in the analysis. You can add normalized box plots, summary tables, capability indices, and an individual report for each variable in the analysis. Specification limits should be entered or available for each variable in the data table.

The individual detail reports include indices that help demonstrate a system's ability to meet a set of requirements. Capability indices help change the focus from just meeting qualifications to continuous improvement of a process.

Capability analysis methods usually require that the data being studied is statistically stable, with no variation present. If a system is stable, capability analysis demonstrates the ability of the system in the past, and can also predict the system's future performance.

The Capability platform has been replaced by the Process Capability platform. See "Process Capability" on page 213 for more information about the new platform. The Capability platform has been removed from the **Analyze > Quality and Process** menu, and it is available only via scripting.

Note: The Process Capability platform expands significantly on the Capability analyses that are available through **Analyze > Distribution** and through **Analyze > Quality and Process > Control Chart**. By default, the Process Capability platform uses the appropriate AIAG notation for capability indices. It denotes an index constructed from an overall variation estimate using Ppk labeling. The previous Capability platform reported only overall variation with Cpk or Ppk labeling based on a platform preference. In the new platform, Cpk corresponds to within variation and Ppk corresponds to overall variation.

Launch the Capability Platform

Launch the Capability platform by submitting the following JSL:

```
Capability();
```

Figure 11.2 The Capability Launch Window for Semiconductor Capability.jmp

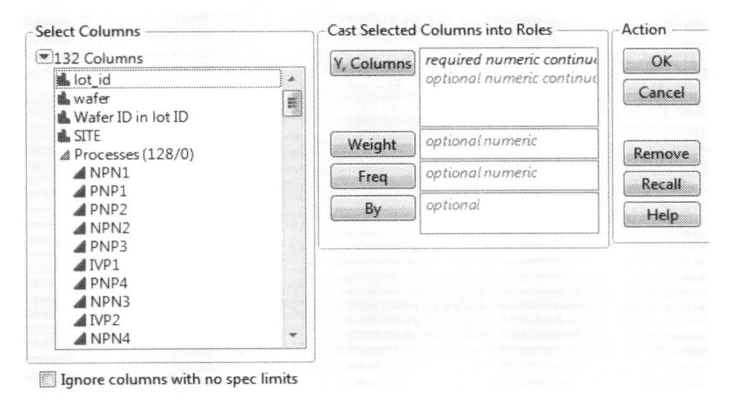

The Capability launch window contains the following options:

Y, Columns Assigns the variables that you want to analyze.

Weight Assigns a variable to give the observations different weights.

Freq Assigns a frequency variable to this role. This is useful if you have summarized data.

By Produces a separate report for each level of the **By** variable. If more than one **By** variable is assigned, a separate report is produced for each possible combination of the levels of the **By** variables.

Ignore columns with no spec limits Prevents the Spec Limits window from appearing when the data table does not include spec limits for a selected variable.

After clicking **OK**, the Spec Limits window appears if you have not specified limits for the selected variables and **Ignore columns with no spec limits** is not selected. Otherwise, the report window appears.

Entering Specification Limits

The lower specification limit (LSL), upper specification limit (USL), and target define the lower bound, upper bound, and target value for a quality process.

There are several ways to enter spec limits:

- Enter limits in the Spec Limits window after selecting columns in the launch window. See "Spec Limits Window" on page 276.
- Import limits from a JMP data table (known as a Limits Table). See "Limits Data Table" on page 276.
- Enter limits as Spec Limits column properties in the data table. See "Spec Limits Column Property" on page 277.

Only one spec limit is required for a selected column. If only the upper spec limit (USL) is specified, the box plot is colored blue. If only the lower spec limit (LSL) is specified, the box plot is colored red.

Spec Limits Window

After you click **OK** on the launch window, the Spec Limits window appears if the data table does not contain limits for the selected columns and you did not select **Ignore columns with no spec limits** on the launch window. Figure 11.3 shows the Spec Limits window for the Cities.jmp sample data table after selecting OZONE, CO, SO2, and NO in the launch window. Enter the known spec limits and click **Next** to view the reports.

Figure 11.3 Spec Limits Window for Cities.jmp

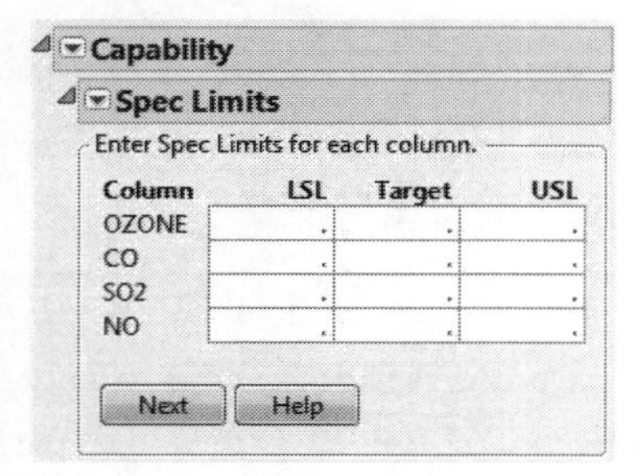

Limits Data Table

You can also retrieve a limits data table with the **Import Spec Limits** option from the Spec Limits red triangle menu. Select **Import Spec Limits** and then select the appropriate data table that contains the specification limits for the analysis. After selecting the appropriate limits table, the values populate the window. Click **Next** to continue to the reports window.

All limits data tables must have a column for each of the variables whose values are the known standard parameters or limits. A limits data table can be in two different formats: *wide* or *tall*. Either you have a column for each response with the limits key or you have one column for each response and the limits key words are the columns. Either of these formats can be read using the **Import Spec Limits** command. Figure 11.4 shows an example of both types of limits data tables.

Figure 11.4 Wide (Top) and Tall (Bottom) Spec Limit Tables

	_LimitsKey	Max deg. F Jan	OZONE	CO	SO2
1	_LSL	15	0	•	0
2	_Target	40	0.1	7	0.05
3	_USL	•	0.3	20	0.1

	Column 1	_LSL	_Target	_USL
1	OZONE	0.075	0.15	0.25
2	CO	5	7	12
3	SO2	0.01	0.04	0.09
4	NO	0.01	0.025	0.04

- A wide table contains one column for each column analyzed in the Capability platform, with three rows plus a _LimitsKey column. In the _LimitsKey column, the three rows need to contain the identifiers _LSL, _USL, and _Target.
- A tall table contains one row for each column analyzed in the Capability platform, with four columns. The first column holds the column names. The other three columns need to be named, _LSL, _USL, and _Target.

The easiest way to create a limits data table is to save results computed by the Capability platform. The **Save Spec Limits** command in the Capability red triangle menu automatically saves limits from the sample values. After entering or loading the specification limits, you can save the limits in the following ways:

- Save the limits to the data table by selecting **Save Spec Limits as Column Properties**.
- Save the limits to a new wide data table, with a column for each analyzed column, by selecting **Save Spec Limits in New Table**.
- Save limits to a new tall data table, with a row for each analyzed column, by selecting **Save Spec Limits in New Table - Tall**. Consider this option for tables that have hundreds of variables.

For more information, see "Capability Platform Options" on page 282.

Spec Limits Column Property

When you perform a capability analysis, you do not have to re-specify the limits each time. Use the **Column Properties > Spec Limits** property in a data table to save specification limits as a column property. Saving these limits in a column also facilitates consistency from use to use. For example, you might run an analysis that uses these limits. When you come back later and change the data, you can run a new analysis on the new data using the same limits. Figure 11.5 displays the Spec Limits section of the Column Properties window for OZONE in the sample data table Cities.jmp.

Enter a lower and upper specification limit and a target value for a numeric column. Ideally, the process variation is between the LSL and USL.

The **Show as graph reference lines** option draws specification limits as reference lines on a graph. For more information about column properties, refer to the Column Info Window chapter in the *Using JMP* book.

Figure 11.5 Spec Limits Section of the Column Properties Window

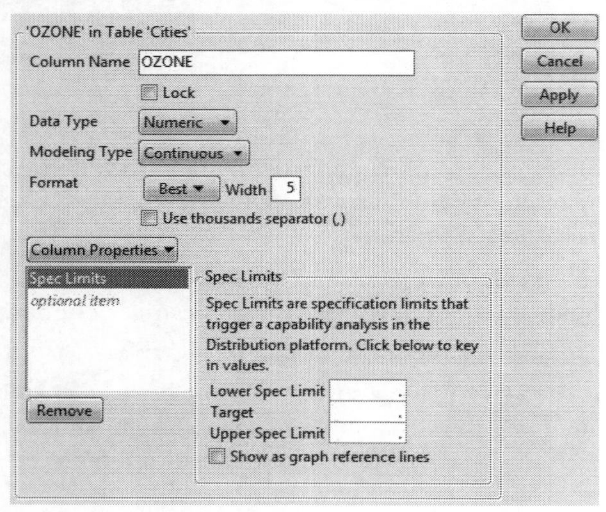

The Capability Report

By default, the Capability report contains two main sections:

- "Goal Plot" on page 279
- "Capability Box Plots" on page 281

Using the **Capability** red triangle menu, you can add normalized box plots, capability indices, and a summary table, as well as display a capability report for each individual variable in the analysis. These options are described in "Capability Platform Options" on page 282.

You can change the report default preference at **File** > **Preferences** > **Platforms** > **Capability**.

Figure 11.6 Default Results for Semiconductor Capability.jmp

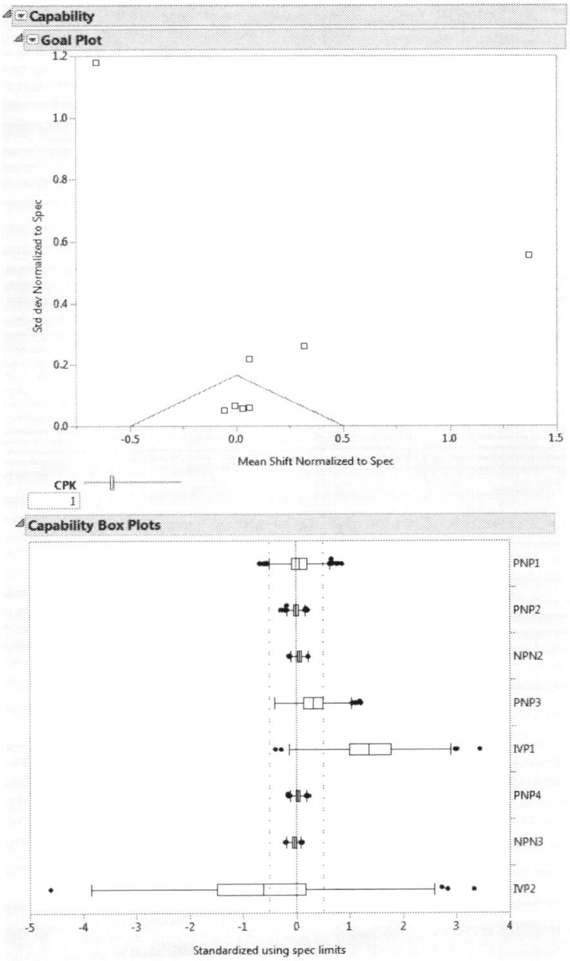

Goal Plot

The Goal Plot shows, for each variable, the spec-normalized mean shift on the *x*-axis, and the spec-normalized standard deviation on the *y*-axis. It is useful for getting a quick, summary view of how the variables are conforming to specification limits. Hold your cursor over each point to view the variable name. The **Goal Plot** red triangle menu has the following commands:

Shade CPK Levels Shows or hides the CPK level shading.

Goal Plot Labels Shows or hides the labels on the points.

Defect Rate Contour Shows or hides a contour representing a specified defect rate.

Figure 11.7 shows the Goal Plot for the entire data set for the Semiconductor Capability.jmp sample data table after selecting **Shade CPK Levels** from the Goal Plot red triangle menu.

Figure 11.7 Goal Plot

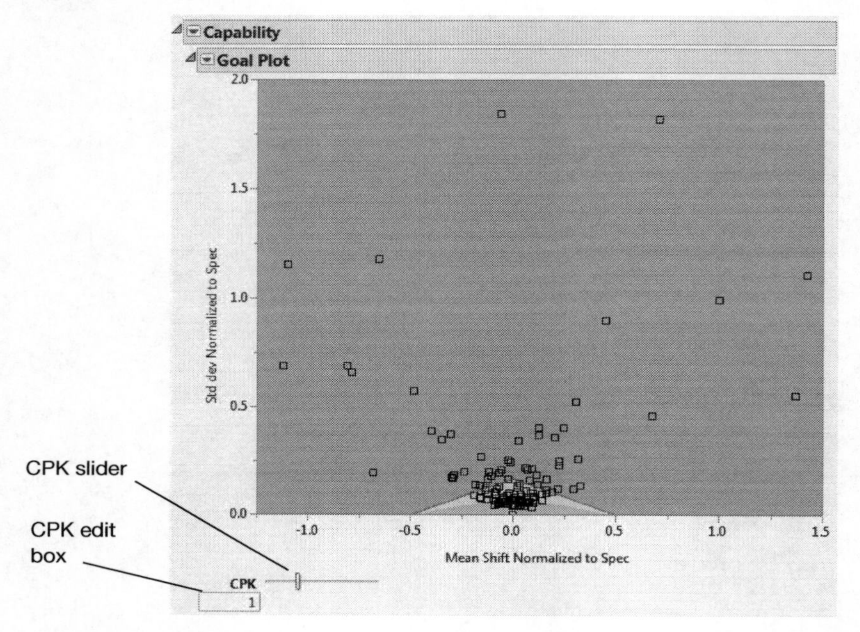

For each column with LSL, Target, and USL, these quantities are defined as follows:

Mean Shift Normalized to Spec = (Mean(Col[*i*]) - Target) / (USL[*i*] - LSL[*i*])

Standard Deviation Normalized to Spec = Standard Deviation(Col[*i*])/ (USL[*i*] - LSL[*i*])

By default, the CPK slider and number edit box is set to CPK = 1. This approximates a non-conformance rate of 0.0027, if the distribution is centered. The red goal line represents the CPK shown in the edit box. To change the CPK value, move the slider or enter a number in the edit box. Points on the plot represent columns, not rows.

The shaded areas are described as follows, with C representing the value shown in the CPK edit box:

- Points in the red area have CPK < C.
- Points in the yellow area have C < CPK < 2C.
- Points in the green area have 2C < CPK.

JMP has a preference for plotting PPK instead of CPK. When this is on, the slider is labeled with PPK. You can change the preference at **File** > **Preferences** > **Platforms** > **Capability**.

Capability Box Plots

The Capability Box Plots show a box plot for each variable selected in the analysis. The values for each column are centered by their target value and scaled by the difference between the specification limits. For each column Y_j:

$$Z_{ij} = \frac{Y_{ij} - T_j}{\text{USL}_j - \text{LSL}_j} \quad \text{with } T_j \text{ being the target}$$

Figure 11.8 Capability Box Plot

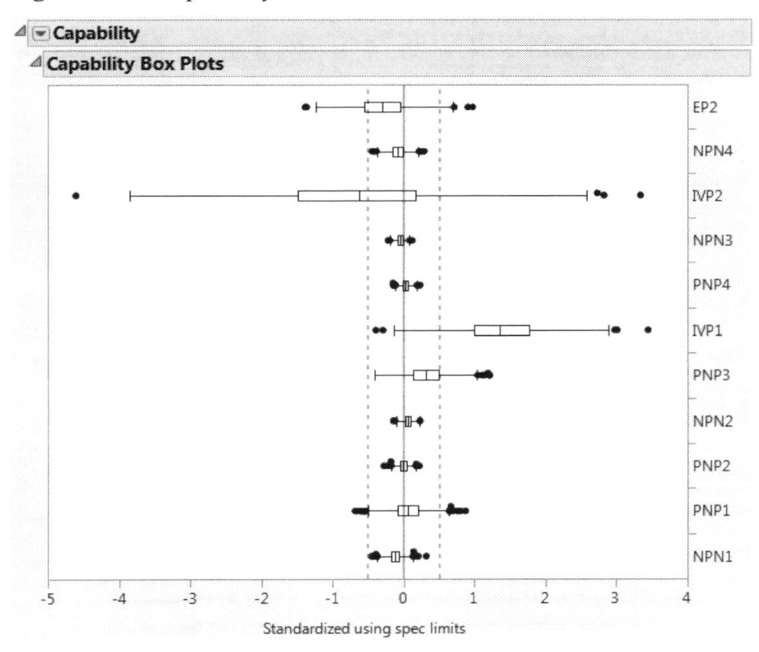

Figure 11.8 shows the Capability Box Plot for the Semiconductor Capability.jmp sample data table. The left and right dotted green lines, drawn at ±0.5, represent the standardized LSL_j and USL_j respectively. This plot is useful for comparing variables with respect to their specification limits. For example, the majority of points for IVP1 are above its USL, while the majority of its points for IVP2 are less than its target. PNP2 looks to be on target with all data points in the spec limits.

Missing Spec Limits

When a spec limit is missing for a column, the box plot is colored. If only the upper spec limit (USL) is specified, the box plot is colored blue. If only the lower spec limit (LSL) is specified, the box plot is colored red. See Figure 11.9. A note at the bottom of the plot describes the calculations used for the plot.

Figure 11.9 Missing Spec Limits

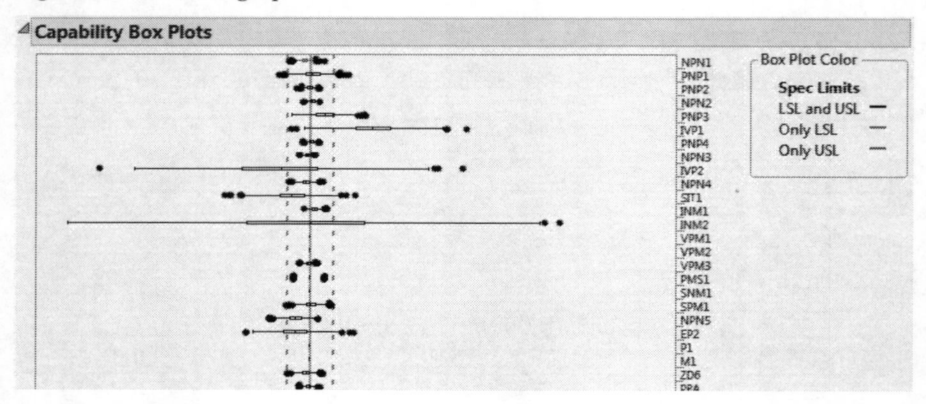

Capability Platform Options

The following options are accessed by clicking the Capability red triangle menu in the report window:

Goal Plot Shows a goal plot for the data. The Goal Plot shows the spec-normalized mean shift on the *x*-axis, and the spec-normalized standard deviation on the *y*-axis for each variable. See "Goal Plot" on page 279 for more information.

Normalized Box Plots Shows a normalized box plot for the data. Each column is standardized by subtracting off the mean and dividing by the standard deviation, and quantiles are formed for each standardized column. The box plots are formed for each column from these standardized quantiles. See "Normalized Box Plots" on page 283 for more information.

Capability Box Plots Shows a capability box plot for each variable in the analysis. The values for each column are centered by their target value and scaled by the difference between the specification limits. See "Capability Box Plots" on page 281 for more information.

Make Summary Table Makes a summary table for the data that includes the variable's name, its spec-normalized mean shift, and its spec-normalized standard deviation. See "Make Summary Table" on page 284 for more information.

Capability Indices Report Shows a capability indices report for the data showing each variable's LSL, Target, USL, Mean, Standard Deviation, Cp, CPK, and PPM. See "Capability Indices Report" on page 285 for more information.

Individual Detail Reports Shows individual detail reports for each variable in the analysis. See "Individual Detail Reports" on page 285 for more information.

Order By Enables you to reorder box plots, capability indices reports, and individual details reports. You can reorder by **Initial Order**, **Reverse Initial Order**, **CPK Ascending**, or **CPK Descending**.

Save Spec Limits Enables you to save specification limits for the data. See "Limits Data Table" on page 276.

See the JMP Reports chapter in the *Using JMP* book for more information about the following options:

Local Data Filter Shows or hides the local data filter that enables you to filter the data used in a specific report.

Redo Contains options that enable you to repeat or relaunch the analysis. In platforms that support the feature, the Automatic Recalc option immediately reflects the changes that you make to the data table in the corresponding report window.

Save Script Contains options that enable you to save a script that reproduces the report to several destinations.

Save By-Group Script Contains options that enable you to save a script that reproduces the platform report for all levels of a By variable to several destinations. Available only when a By variable is specified in the launch window.

Normalized Box Plots

When drawing Normalized Box Plots, JMP first standardizes each column by subtracting off the mean and dividing by the standard deviation. Next, quantiles are formed for each standardized column. The box plots are formed for each column from these standardized quantiles.

Figure 11.10 Normalized Box Plot

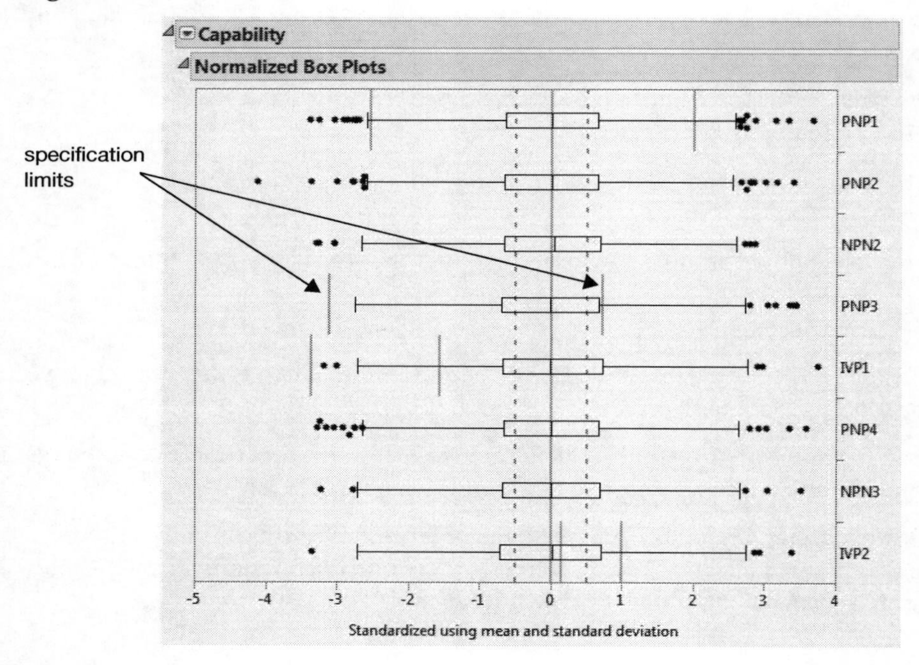

Figure 11.10 shows the Normalized Box Plot for the Semiconductor Capability.jmp sample data table.

The green vertical lines represent the spec limits normalized by the mean and standard deviation. The gray dotted vertical lines are drawn at ±0.5, since the data is standardized to a standard deviation of 1.

Make Summary Table

The Make Summary Table option produces a summary table that includes the variable's name, its spec-normalized mean shift, and its spec-normalized standard deviation for further evaluation.

Figure 11.11 shows the Summary Table for the Semiconductor Capability.jmp sample data table.

Figure 11.11 Summary Table

	Variable	Mean Shift Normalized to Spec	Std dev Normalized to Spec
1	PNP1	0.060514123	0.2199887772
2	PNP2	-0.007336131	0.0663486129
3	NPN2	0.0580913305	0.0612288463
4	PNP3	0.3153977755	0.2609703636
5	IVP1	1.368139782	0.5535989822
6	PNP4	0.0300797696	0.0556116319
7	NPN3	-0.056828799	0.0503418504
8	IVP2	-0.654156076	1.1799901131

Capability Indices Report

The Capability Indices Report option shows or hides a table showing each variable's LSL, Target, USL, Mean, Standard Deviation, Cp, CPK, and PPM. Figure 11.12 shows the Capability Indices for the Semiconductor Capability.jmp sample data table. Optional columns for this report include Lower CI, Upper CI, CPM, CPL, CPU, PPM Below LSL, and PPM Above USL, among others. To reveal these optional columns, right-click on the report and select the column names from the **Columns** submenu.

Figure 11.12 Capability Indices

Capability

Capability Indices

Columns	LSL	Target	USL	Mean	Standard Deviation	CP	CPK	PPM
PNP1	164.3895	297.0179	429.6463	313.0697	58.35353	0.757614	0.6659	28289.42
PNP2	-136.122	465.442	1067.006	456.6157	79.82589	2.511984	2.4751	6.655e-8
NPN2	96.59381	113.749	130.9042	115.7421	2.100786	2.722029	2.4058	2.651e-7
PNP3	118.6778	130.2898	141.9018	137.6146	6.060762	0.638642	0.2358	240559.3
IVP1	59.62007	63.41011	67.20015	73.78072	4.196326	0.30106	-0.5227	941949.5
PNP4	-54.4319	238.7386	531.9091	256.3756	32.60738	2.996975	2.8167	7.73e-16
NPN3	97.31768	120.8047	144.2917	118.1352	2.364757	3.310698	2.9344	6.65e-13
IVP2	139.2004	142.3052	145.4099	138.2432	7.327164	0.141244	-0.0435	715981.5

Individual Detail Reports

The **Individual Detail Reports** option displays a capability report for each variable in the analysis. This report is identical to the Capability Analysis Report from the Distribution platform, detailed in the Distributions chapter in the *Basic Analysis* book. Figure 11.13 shows the Individual Detail Report for PNP1 from the Semiconductor Capability.jmp sample data table.

Figure 11.13 Individual Detail Report

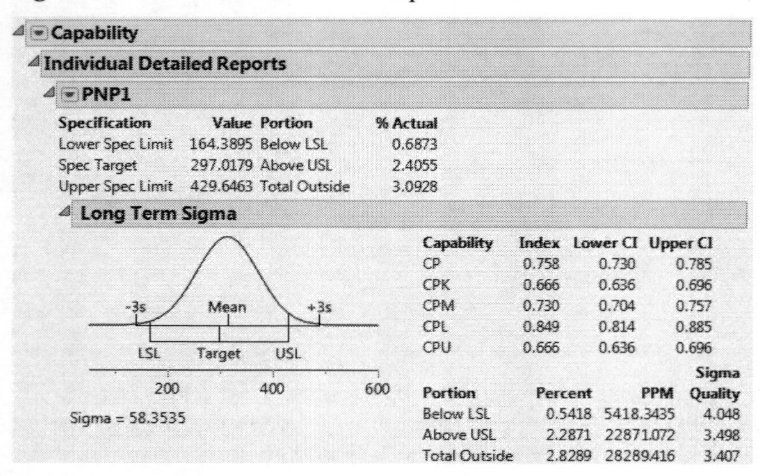

Specification	Value	Portion	% Actual
Lower Spec Limit	164.3895	Below LSL	0.6873
Spec Target	297.0179	Above USL	2.4055
Upper Spec Limit	429.6463	Total Outside	3.0928

Capability	Index	Lower CI	Upper CI
CP	0.758	0.730	0.785
CPK	0.666	0.636	0.696
CPM	0.730	0.704	0.757
CPL	0.849	0.814	0.885
CPU	0.666	0.636	0.696

Portion	Percent	PPM	Sigma Quality
Below LSL	0.5418	5418.3435	4.048
Above USL	2.2871	22871.072	3.498
Total Outside	2.8289	28289.416	3.407

Chapter 12

Pareto Plots

Focus Improvement Efforts on the Vital Few

Improve the statistical quality of your process or operation using Pareto plots. A Pareto plot is a chart that shows severity (frequency) of problems in a quality-related process or operation. Pareto plots help you decide which problems to solve first by highlighting the frequency and severity of problems.

Figure 12.1 Pareto Plot Examples

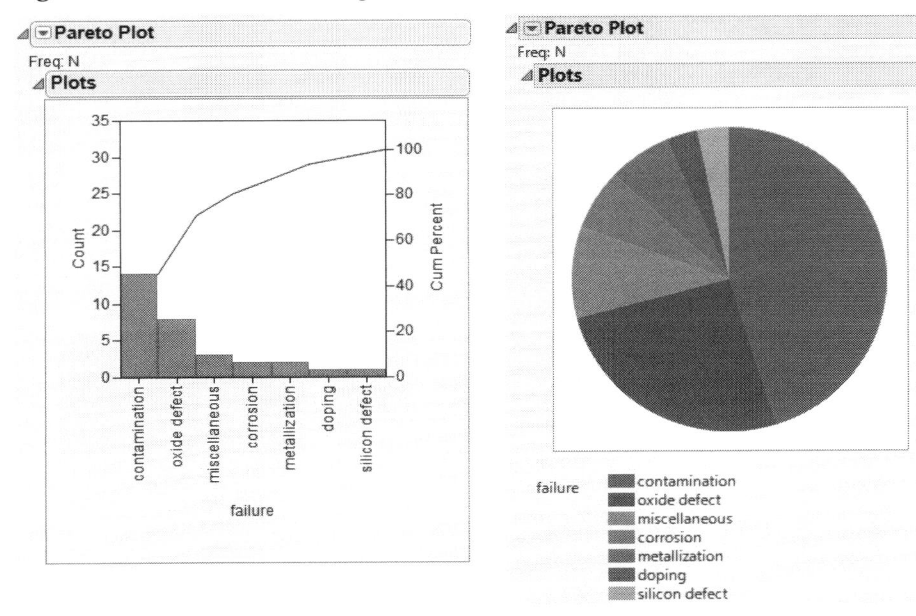

Overview of the Pareto Plot Platform

The Pareto Plot platform produces charts to display the relative frequency or severity of problems in a quality-related process or operation. The Pareto plot is displayed initially as a bar chart that shows the classification of problems arranged in decreasing order. The column whose values are the cause of a problem is assigned the *Y* role and is called the *process variable*.

You can also generate a comparative Pareto plot, which combines two or more Pareto plots for the same process variable. The single display shows plots for each value in a column assigned the *X* role, or combination of levels from two *X* variables. Columns assigned the *X* role are called *classification variables*.

The Pareto plot can chart a single *Y* (process) variable with no *X* classification variables, with a single *X*, or with two *X* variables. The Pareto function does not distinguish between numeric and character variables or between modeling types. You can switch between a bar chart and a pie chart. All values are treated as discrete, and bars or wedges represent either counts or percentages.

Example of the Pareto Plot Platform

This example uses the Failure.jmp sample data table, which contains failure data and a frequency column. It lists causes of failure during the fabrication of integrated circuits and the number of times each type of defect occurred. From the analysis, you can determine which factors contribute most toward process failure.

1. Select **Help > Sample Data Library** and open Quality Control/Failure.jmp.
2. Select **Analyze** > **Quality and Process > Pareto Plot**.
3. Select failure and click **Y, Cause**.

   This column lists the causes of failure and is the variable that you want to inspect.
4. Select N and click **Freq.**

   This column list the number of times each type of defect occurred.
5. Click **OK.**

Figure 12.2 Pareto Plot Report Window

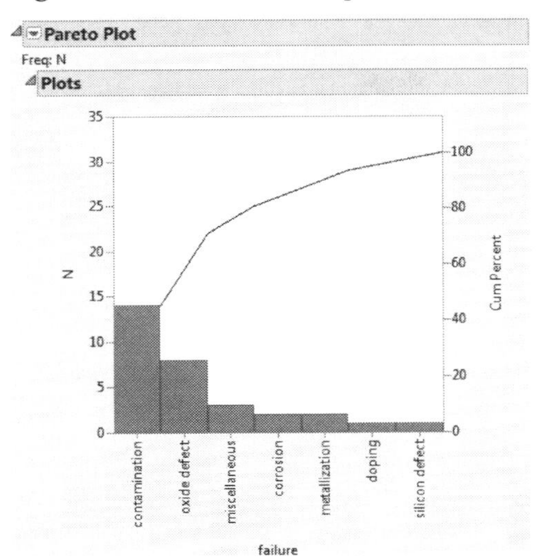

The left axis represents the count of failures, and the right axis represents the percent of failures in each category. The bars are in decreasing order with the most frequently occurring failure to the left. The curve indicates the cumulative failures from left to right.

6. Select **Label Cum Percent Points** from the red triangle menu next to Pareto Plot.

   Note that Contamination accounts for approximately 45% of the failures. The point above the Oxide Defect bar shows that Contamination and Oxide Defect together account for approximately 71% of the failures.

7. From the red triangle menu, deselect **Label Cum Percent Points** and **Show Cum Percent Curve.**
8. Double-click the y-axis labeled **N** and rename it **Count**.
9. Double-click the y-axis to display the **Y Axis Specification** window.
   - In the **Maximum** field, type 15.
   - In the **Increment** field, type 2.
   - In the **Tick Marks and Grid Lines** area, select **Grid Lines** for the **Major** grid line.
   - Click **OK.**
10. From the red triangle menu, select **Category Legend.**

Figure 12.3 Pareto Plot with Display Options

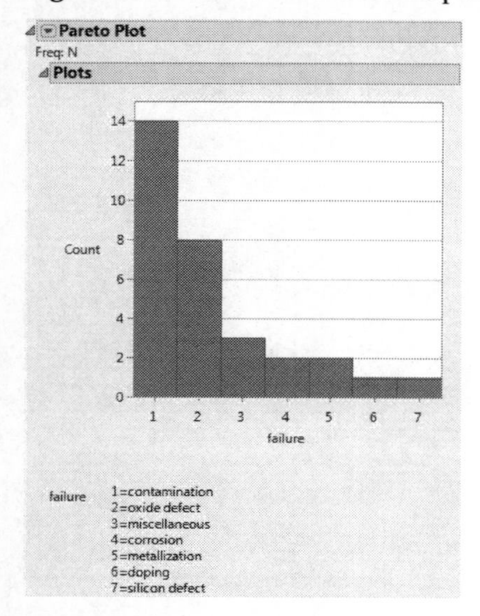

Figure 12.3 shows counts now instead of percents, and has a category legend. The vertical count axis is rescaled and has grid lines at the major tick marks.

11. To view the data as a pie chart, select Pie Chart from the red triangle menu.

Figure 12.4 Pareto Plot as a Pie Chart

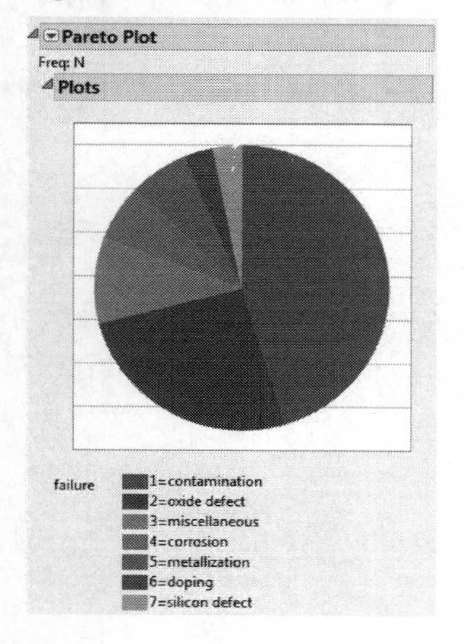

Contamination and Oxide Defect clearly represent the majority of the failures.

Launch the Pareto Plot Platform

Launch the Pareto Plot platform by selecting **Analyze > Quality and Process > Pareto Plot.**

Figure 12.5 The Pareto Plot Launch Window

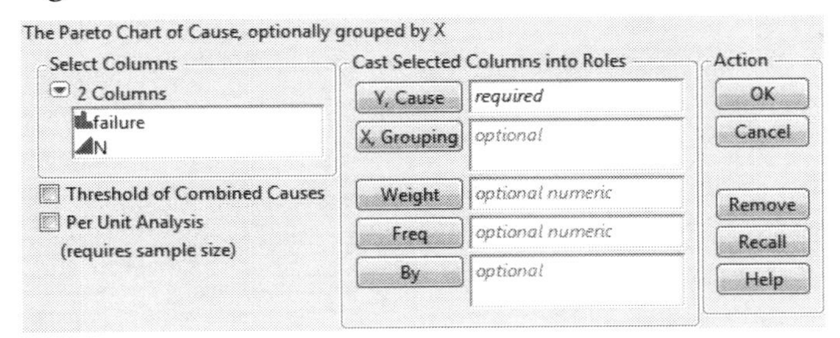

The Pareto Plot launch window contains the following options:

Y, Cause Identifies the column whose values are the cause of a problem. It is called the process variable and is the variable that you want to inspect.

X, Grouping Identifies the grouping factor. The grouping variable produces one Pareto plot window with side-by-side plots for each value. You can have no grouping variable, one grouping variable (see "One-Way Comparative Pareto Plot Example" on page 300), or two grouping variables (see "Two-Way Comparative Pareto Plot Example" on page 301).

Weight Assigns a variable to give the observations different weights.

Freq Identifies the column whose values hold the frequencies.

By Identifies a variable to produce a separate analysis for each value that appears in the column.

Threshold of Combined Causes Enables you to specify a threshold for combining causes by specifying a minimum rate or count. Select the option and then select **Tail %** or **Count** and enter the threshold value. The Tail % option combines smaller count groups against the percentage specified of the total (combined small groups count/total group count). The Count option enables you to specify a specific count threshold. For an example, see "Threshold of Combined Causes Example" on page 295.

Per Unit Analysis Enables you to compare defect rates across groups. JMP calculates the defect rate as well as 95% confidence intervals of the defect rate. Select the option and then select **Constant** or **Value in Freq Column** and enter the sample size value or cause code, respectively. The Constant option enables you to specify a constant sample size on the launch window. The Value In Freq Column option enables you to specify a unique sample size for a group through a special cause code to designate the rows as cause rows.

Although causes are allowed to be combined in Pareto plots, the calculations for these analyses do not change correspondingly.

For examples, see "Using a Constant Size across Groups Example" on page 297 and "Using a Non-Constant Sample Size across Groups Example" on page 298.

The Pareto Plot Report

The Pareto plot combines a bar chart displaying percentages of variables in the data with a line graph showing cumulative percentages of the variables.

Figure 12.6 Pareto Plot Example

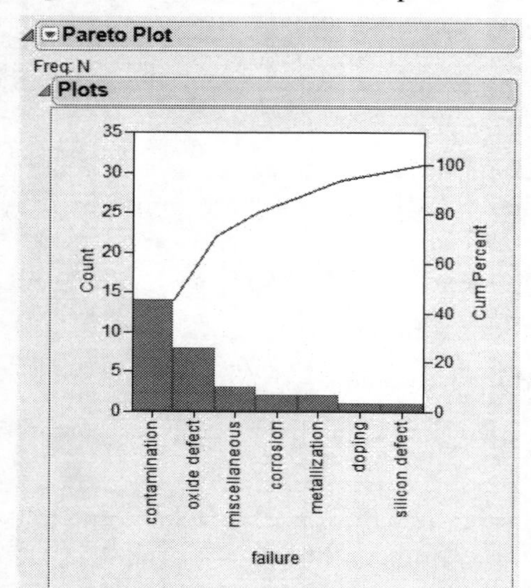

The Pareto plot can chart a single *Y* (process) variable with no *X* classification variables, with a single *X*, or with two *X* variables. The Pareto plot does not distinguish between numeric and character variables or between modeling types. All values are treated as discrete, and bars represent either counts or percentages. The following list describes the arrangement of the Pareto plot:

- A *Y* variable with no *X* classification variables produces a single chart with a bar for each value of the *Y* variable. For an example, see "Example of the Pareto Plot Platform" on page 288.

- A *Y* variable with one *X* classification variable produces a row of Pareto plots. There is a plot for each level of the *X* variable with bars for each *Y* level. These plots are referred to as the *cells* of a comparative Pareto plot. There is a cell for each level of the *X* (classification) variable. Because there is only one *X* variable, this is called a *one-way comparative Pareto plot*. For an example, see "One-Way Comparative Pareto Plot Example" on page 300.
- A *Y* variable with two *X* variables produces rows and columns of Pareto plots. There is a row for each level of the first *X* variable and a column for each level of the second *X* variable. Because there are two *X* variables, this is called a *two-way comparative Pareto plot*. The rows have a Pareto plot for each value of the first *X* variable, as described previously. The upper left cell is called the *key cell*. Its bars are arranged in descending order. The bars in the other cells are in the same order as the key cell. You can reorder the rows and columns of cells. The cell that moves to the upper left corner becomes the new key cell and the bars in all other cells rearrange accordingly. For an example, see "Two-Way Comparative Pareto Plot Example" on page 301.
- Each bar is the color for which the rows for that *Y* level are assigned in the associated data table. Otherwise, a single color is used for all of the bars whose *Y* levels do not have rows with an assigned color. If the rows for a *Y* level have different colors, the bar for that *Y* level is the color of the first row for that *Y* level in the data table.

You can change the type of scale and arrangement of bars and convert the bars into a pie chart using the options in the Pareto Plot red triangle menu. For more information, see "Pareto Plot Platform Options" on page 293.

Pareto Plot Platform Options

The red triangle menu next to Pareto Plot has commands that customize the appearance of the plots. It also has options in the **Causes** submenu that affect individual bars within a Pareto plot. The following commands affect the appearance of the Pareto plot as a whole:

Percent Scale Shows or hides the count and percent left vertical axis display.

N Legend Shows or hides the total sample size in the plot area.

Category Legend Shows or hides labeled bars and a separate category legend.

Pie Chart Shows or hides the bar chart and pie chart representation.

Reorder Horizontal, Reorder Vertical Reorders grouped Pareto plots when there is one or more grouping variables.

Ungroup Plots Splits up a group of Pareto charts into separate plots.

Count Analysis Performs defect per unit analyses. Enables you to compare defect rates and perform ratio tests across and within groups:

- **Per Unit Rates** compares defect rates across groups. If a sample size is specified, Defects Per Unit (DPU) and Parts Per Million (PPM) columns are added to the report.
- **Test Rate Within Groups** tests (a likelihood ratio chi-square) whether the Defects Per Unit (DPU) across causes are the same within a group.
- **Test Rates Across Groups** tests (a likelihood ratio chi-square) whether the Defects Per Unit (DPU) for each cause is the same across groups.

Show Cum Percent Curve Shows or hides the cumulative percent curve above the bars and the cumulative percent axis on the vertical right axis.

Show Cum Percent Axis Shows or hides the cumulative percent axis on the vertical right axis.

Show Cum Percent Points Shows or hides the points on the cumulative percent curve.

Label Cum Percent Points Shows or hides the labels on the points on the cumulative curve.

Cum Percent Curve Color Changes the color of the cumulative percent curve.

Causes Has options that affect one or more individual chart bars. See "Causes Options" on page 294, for a description of these options.

See the JMP Reports chapter in the *Using JMP* book for more information about the following options:

Local Data Filter Shows or hides the local data filter that enables you to filter the data used in a specific report.

Redo Contains options that enable you to repeat or relaunch the analysis. In platforms that support the feature, the Automatic Recalc option immediately reflects the changes that you make to the data table in the corresponding report window.

Save Script Contains options that enable you to save a script that reproduces the report to several destinations.

Save By-Group Script Contains options that enable you to save a script that reproduces the platform report for all levels of a By variable to several destinations. Available only when a By variable is specified in the launch window.

Causes Options

You can highlight a bar by clicking on it. Use Control-click to select multiple bars that are not contiguous. When you select bars, you can access the commands on the red triangle menu that affect Pareto plot bars. They are found on the **Causes** submenu on the red triangle menu. These options are also available with a right-click anywhere in the plot area. The following options apply to highlighted bars instead of to the chart as a whole:

Combine Causes Combines selected (highlighted) bars. You can select either **Selected, Last Causes, First Causes** or select from a list of variables as shown in Figure 12.7.

Figure 12.7 Combine Causes Window

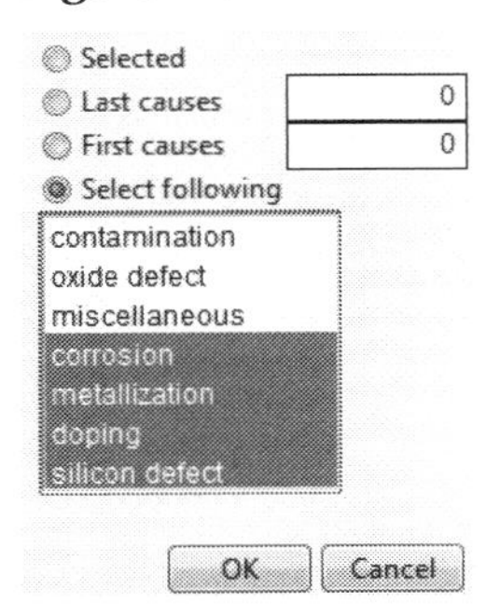

Separate Causes Separates selected bars into their original component bars.

Move to First Moves one or more highlighted bars to the left (first) position.

Move to Last Moves one or more highlighted bars to the right (last) position.

Colors Shows the colors palette for coloring one or more highlighted bars.

Markers Shows the markers palette for assigning a marker to the points on the cumulative percent curve, when the **Show Cum Percent Points** command is in effect.

Label Displays the bar value at the top of all highlighted bars.

Additional Examples of the Pareto Plot Platform

This section contains additional examples using the Pareto Plot platform.

Threshold of Combined Causes Example

This example uses the Failure.jmp sample data table, which contains failure data and a frequency column. It lists causes of failure during the fabrication of integrated circuits and the number of times each type of defect occurred. A threshold value of 2 is specified for this example.

1. Select **Help > Sample Data Library** and open Quality Control/Failure.jmp.
2. Select **Analyze** > **Quality and Process > Pareto Plot.**
3. Select failure and click **Y, Cause.**
4. Select N and click **Freq.**
5. Select **Threshold of Combined Causes** and then select **Count.**
6. Enter 2 as the threshold value.
7. Click **OK.**

Figure 12.8 Pareto Plot with a Threshold Count of 2

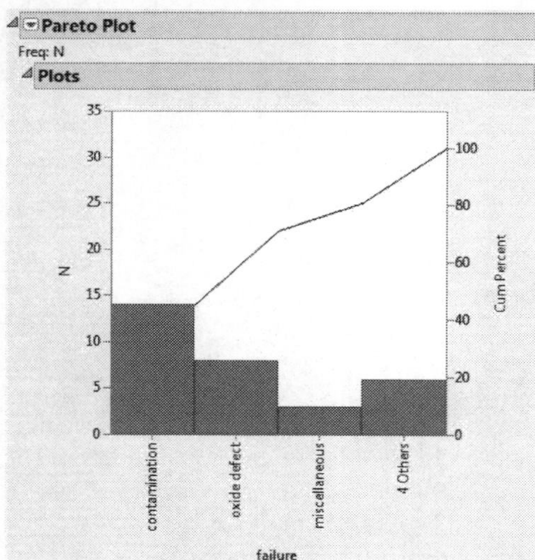

Figure 12.8 displays the plot after specifying a count of 2. All causes with counts 2 or fewer are combined into the final bar labeled 4 Others.

8. To separate the combined bars into original categories as shown in Figure 12.9, select **Causes** > **Separate Causes**.

Figure 12.9 Pareto Plot with Separated Causes

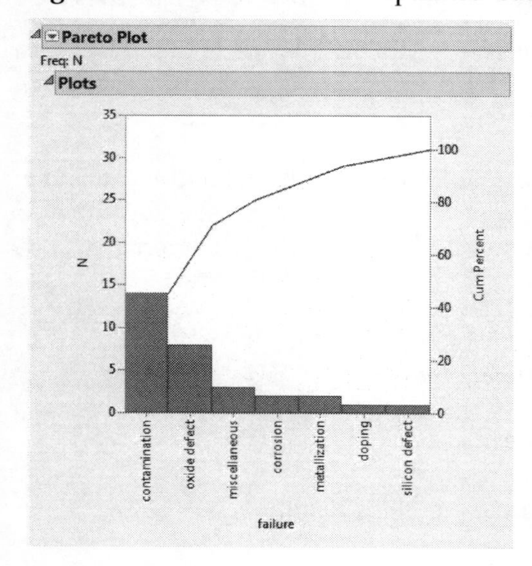

Using a Constant Size across Groups Example

This example uses the Failures.jmp sample data table, which contains failure data and a frequency column. It lists causes of failure during the fabrication of integrated circuits and the number of times each type of defect occurred for two processes. A constant sample size of 1000 is specified for this example.

1. Select **Help > Sample Data Library** and open Quality Control/Failures.jmp.
2. Select **Analyze** > **Quality and Process > Pareto Plot.**
3. Select Causes and click **Y, Cause**.
4. Select Process and click **X, Grouping.**
5. Select Count and click **Freq.**
6. Select **Per Unit Analysis** and then select **Constant.**
7. Enter *1000* in **Sample Size.**
8. Click **OK.**

Figure 12.10 Pareto Plot Report Window

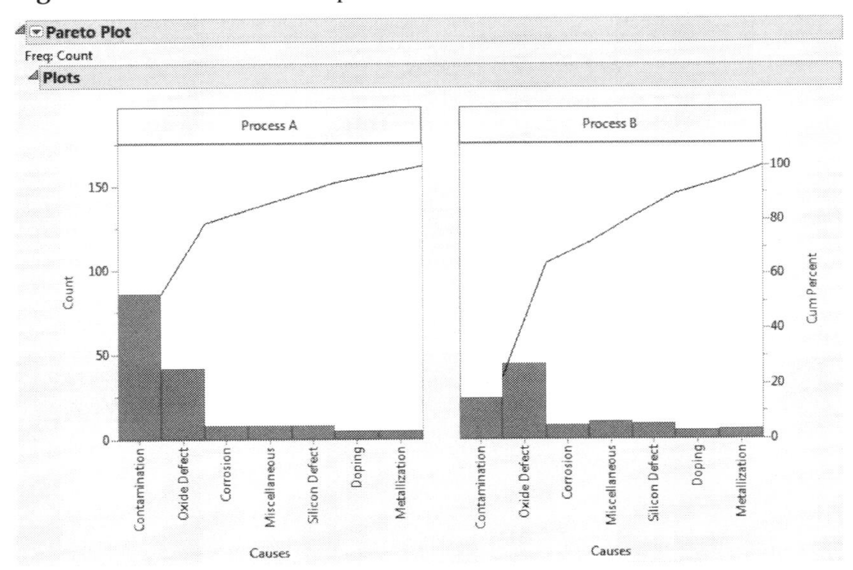

Process A indicates Contamination as the top failure while Process B indicates Oxide Defect as the leading failure.

9. Select **Count Analysis** > **Test Rates Across Groups** from the red triangle menu.

Figure 12.11 Test Rates across Groups Results

Test Rate Across Groups

Test rate across group: Process

Cause	DPU Diff	-.8 -.6 -.4 -.2 0 .2 .4 .6 .8	Std Error	ChiSquare	DF	Prob>ChiSq
Contamination	0.0620		0.0105	37.0810	1	<.0001*
Oxide Defect	-0.0020		0.0093	0.0465	1	0.8292
Corrosion	0.0000		0.0040	0.0000	1	1.0000
Miscellaneous	-0.0020		0.0042	0.2227	1	0.6370
Silicon Defect	-0.0010		0.0041	0.0589	1	0.8083
Doping	0.0000		0.0032	0.0000	1	1.0000
Metallization	-0.0010		0.0033	0.0910	1	0.7629
Pooled Total	0.0080		0.0023	11.7882	1	0.0006*

Note that the DPU for Contamination across groups (Process A and B) is around 0.06.

Using a Non-Constant Sample Size across Groups Example

This example uses the Failuressize.jmp sample data table, which contains failure data and a frequency column. It lists causes of failure during the fabrication of integrated circuits and the number of times each type of defect occurred for two processes. Among the other causes (Oxide Defect, Silicon Defect, and so on) is a cause labeled *size*. Specifying size as the cause code designates the rows as size rows.

1. Select **Help > Sample Data Library** and open Quality Control/Failuressize.jmp.
2. Select **Analyze** > **Quality and Process > Pareto Plot.**
3. Select Causes and click **Y, Cause.**
4. Select Process and click **X, Grouping.**
5. Select Count and click **Freq.**
6. Select **Per Unit Analysis** and then select **Value in Freq Column.**
7. Enter *size* in **Cause Code.**
8. Click **OK.**

Figure 12.12 Pareto Plot Report Window

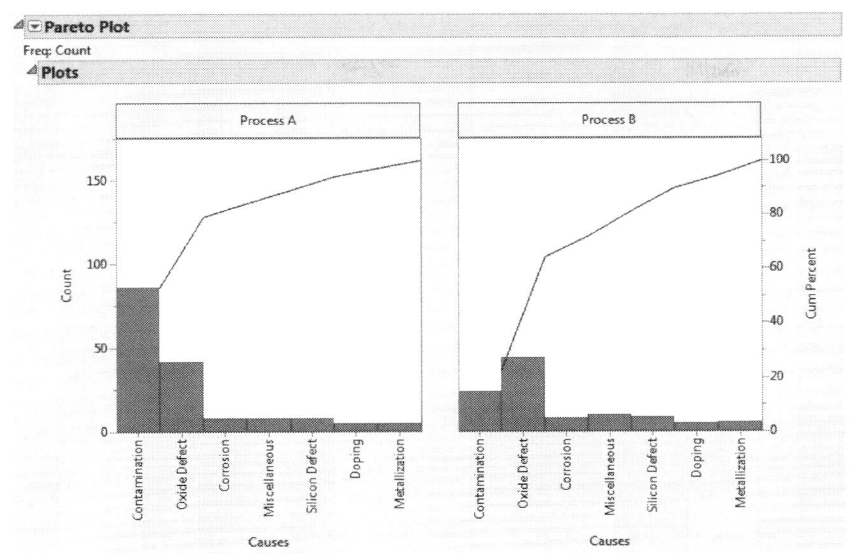

9. Select **Count Analysis** > **Per Unit Rates** and **Count Analysis** > **Test Rates Across Groups** from the red triangle menu.

Figure 12.13 Per Unit Rates and Test Rates across Groups Results

Per Unit Rates

Process	Cause	Count	DPU	PPM	Lower 95%	Upper 95%
Process A	Contamination	86	0.8515	851485.15	0.6811	1.0516
	Oxide Defect	42	0.4158	415841.58	0.2997	0.5621
	Corrosion	8	0.0792	79207.92	0.0342	0.1561
	Miscellaneous	8	0.0792	79207.92	0.0342	0.1561
	Silicon Defect	8	0.0792	79207.92	0.0342	0.1561
	Doping	5	0.0495	49504.95	0.0161	0.1155
	Metallization	5	0.0495	49504.95	0.0161	0.1155
	Pooled Total	162	0.2291	229137.20	0.1952	0.2673
	size	101				
Process B	Contamination	24	0.1655	165517.24	0.1061	0.2463
	Oxide Defect	44	0.3034	303448.28	0.2205	0.4074
	Corrosion	8	0.0552	55172.41	0.0238	0.1087
	Miscellaneous	10	0.0690	68965.52	0.0331	0.1268
	Silicon Defect	9	0.0621	62068.97	0.0284	0.1178
	Doping	5	0.0345	34482.76	0.0112	0.0805
	Metallization	6	0.0414	41379.31	0.0152	0.0901
	Pooled Total	106	0.1044	104433.50	0.0855	0.1263
	size	145				

Test Rate Across Groups

Test rate across group: Process

Cause	DPU Diff	-.8-.6-.4-.2 0 .2 .4 .6 .8	Std Error	ChiSquare	DF	Prob>ChiSq
Contamination	0.6860		0.0978	63.0776	1	<.0001*
Oxide Defect	0.1124		0.0788	2.1195	1	0.1454
Corrosion	0.0240		0.0341	0.5202	1	0.4707
Miscellaneous	0.0102		0.0355	0.0847	1	0.7710
Silicon Defect	0.0171		0.0348	0.2500	1	0.6171
Doping	0.0150		0.0270	0.3251	1	0.5685
Metallization	0.0081		0.0278	0.0871	1	0.7679
Pooled Total	0.1247		0.0207	40.7524	1	<.0001*

Note that the sample size of 101 is used to calculate the DPU for the causes in group A. However, the sample size of 145 is used to calculate the DPU for the causes in group B.

If there are two group variables (say, Day and Process), Per Unit Rates lists DPU or rates for every combination of Day and Process for each cause. However, Test Rate Across Groups only tests overall differences between groups.

One-Way Comparative Pareto Plot Example

This example uses the Failure2.jmp sample data table. This table records failures in a sample of capacitors manufactured before cleaning a tube in the diffusion furnace and in a sample manufactured after cleaning the furnace. For each type of failure, the variable clean identifies the samples with the values "before" or "after."

1. Select **Help > Sample Data Library** and open Quality Control/Failure2.jmp.
2. Select **Analyze** > **Quality and Process > Pareto Plot.**
3. Select failure and click **Y, Cause.**
4. Select clean and click **X, Grouping.**
5. Select N and click **Freq.**
6. Click **OK.**

Figure 12.14 displays the side-by-side plots for each value of the variable, clean.

Figure 12.14 One-way Comparative Pareto Plot

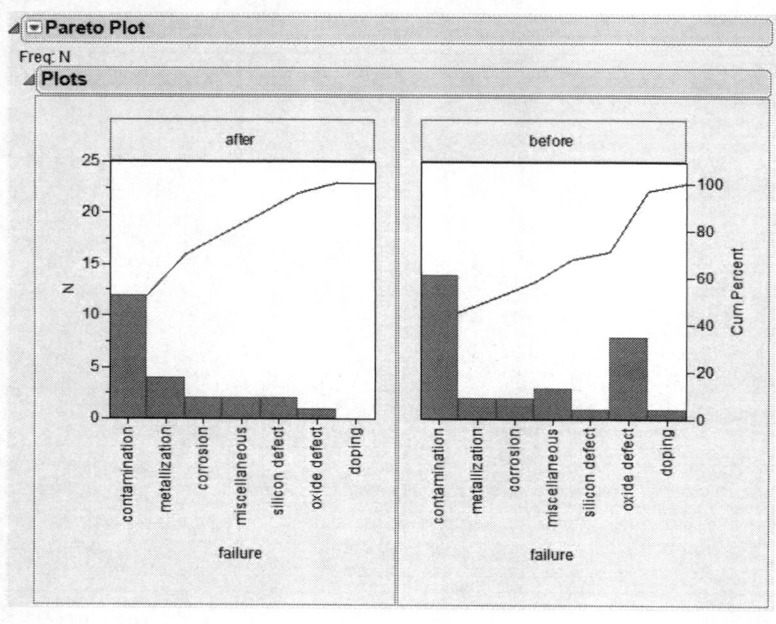

The horizontal and vertical axes are scaled identically for both plots. The bars in the first plot are in descending order of the *y*-axis values and determine the order for all cells.

7. Rearrange the order of the plots by clicking the title (*after*) in the first tile and dragging it to the title of the next tile (*before*).

   A comparison of these two plots shows a reduction in oxide defects after cleaning. However, the plots are easier to interpret when presented as the before-and-after plot shown in Figure 12.15. Note that the order of the causes changes to reflect the order based on the first cell.

Figure 12.15 One-way Comparative Pareto Plot with Reordered Cells

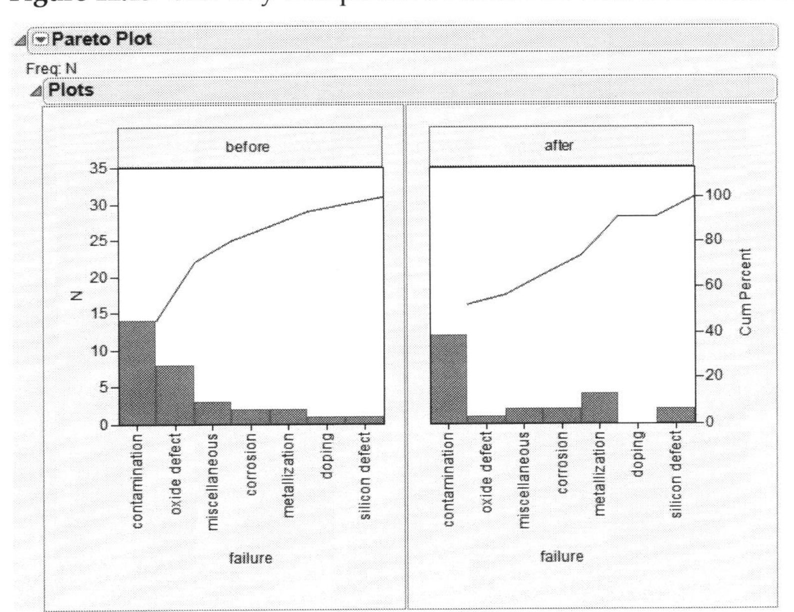

Two-Way Comparative Pareto Plot Example

This example uses the Failure3.jmp sample data table. The data monitors production samples before and after a furnace cleaning for three days for a capacitor manufacturing process. The data table has a column called date with values OCT 1, OCT 2, and OCT 3.

1. Select **Help > Sample Data Library** and open Quality Control/Failure3.jmp.
2. Select **Analyze** > **Quality and Process > Pareto Plot.**
3. Select failure and click **Y, Cause.**
4. Select clean and date and click **X, Grouping.**
5. Select N and click **Freq.**
6. Click **OK.**

Figure 12.16 displays the Pareto plot with a two-way layout of plots that show each level of both X variables. The upper left cell is called the *key cell*. Its bars are arranged in descending order. The bars in the other cells are in the same order as the key cell.

7. Click **Contamination** and **Metallization** in the key cell and the bars for the corresponding categories highlight in all other cells.

Figure 12.16 Two-way Comparative Pareto Plot

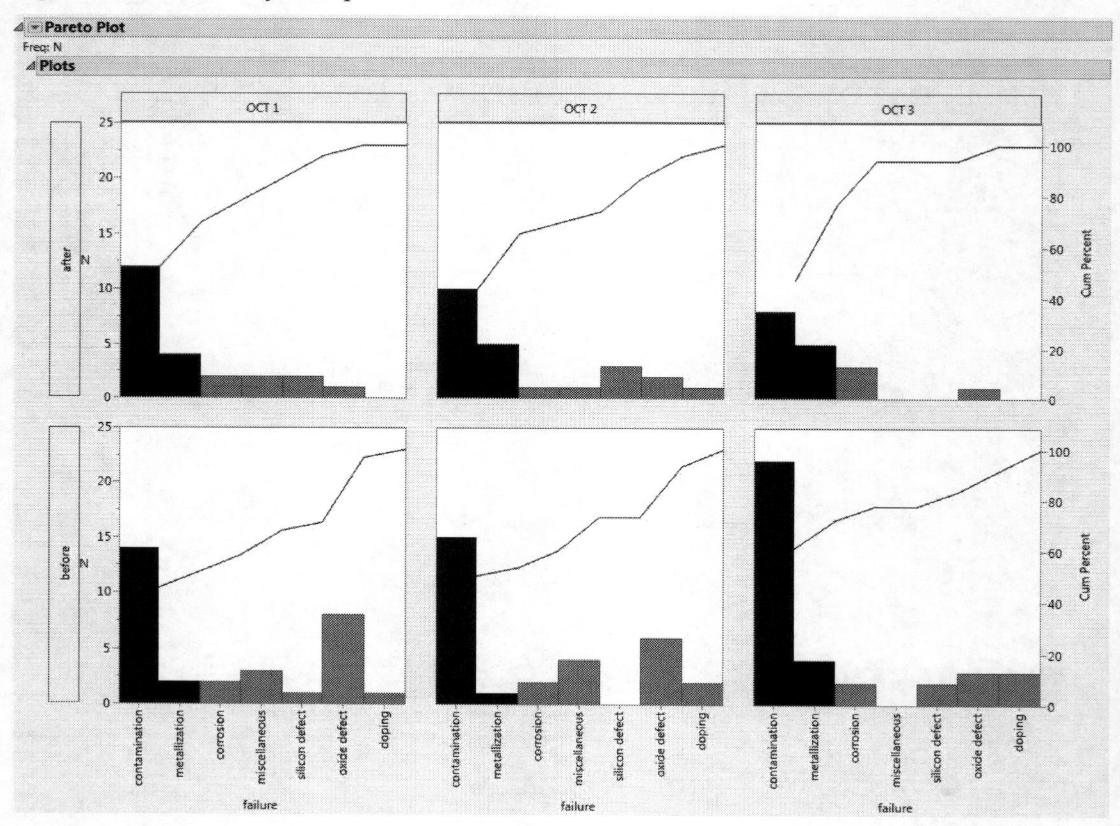

The Pareto plot shown in Figure 12.16 illustrates highlighting the *vital few*. In each cell of the two-way comparative plot, the bars representing the two most frequently occurring problems are selected. Contamination and Metallization are the two vital categories in all cells. After furnace cleaning, Contamination is less of a problem.

Chapter 13

Cause-and-Effect Diagrams

Identify Root Causes

Use the Diagram platform to construct cause-and-effect diagrams, also known as *Ishikawa charts* or *fishbone charts*. Use these diagrams to:

- Organize the causes of an effect (sources of a problem)
- Brainstorm
- Identify variables in preparation for further experimentation

Figure 13.1 Example of a Cause-and-Effect Diagram

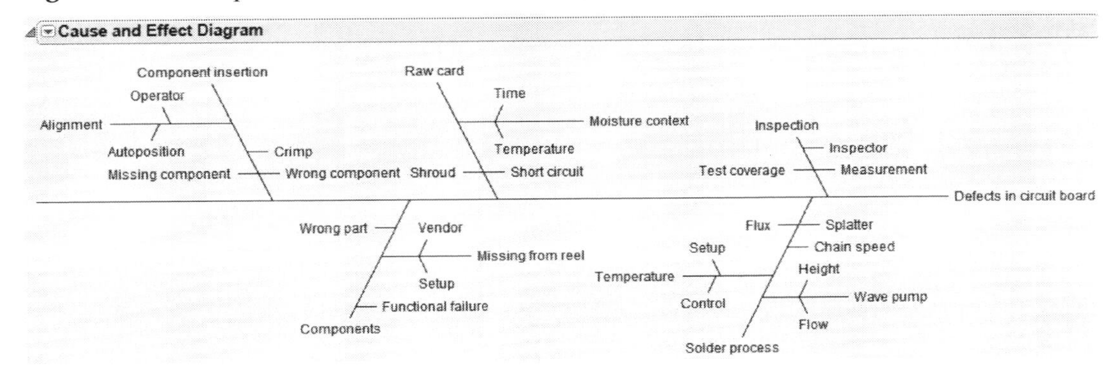

Cause-and-Effect Diagram Overview

Use the Diagram platform to construct cause-and-effect diagrams, also known as *Ishikawa charts* or *fishbone charts*. Use these diagrams to:

- Organize the causes of an effect (sources of a problem)
- Brainstorm
- Identify variables in preparation for further experimentation

Example of a Cause-and-Effect Diagram

You have data about defects in a circuit board. You want to examine the major factors and possible causes of the defects in a diagram.

1. Select **Help > Sample Data Library** and open Ishikawa.jmp.
2. Select **Analyze > Quality and Process > Diagram.**
3. Select Parent and click **X, Parent**.
4. Select Child and click **Y, Child**.
5. Click **OK**.

Figure 13.2 Ishikawa.jmp Diagram

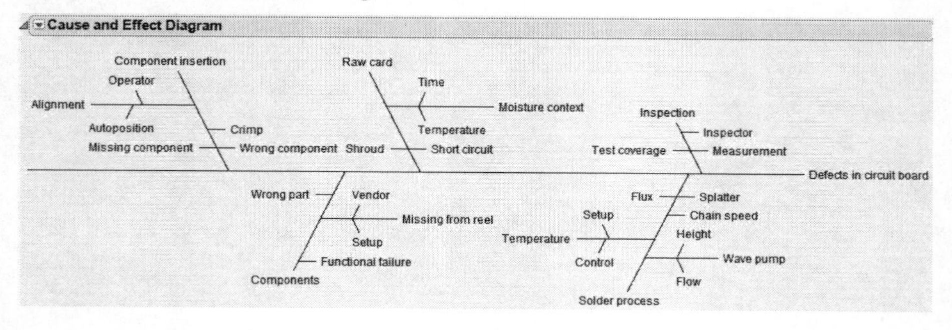

The major factors are Inspection, Solder process, Raw card, Components, and Component insertion. From each major factor, possible causes branch off, such as Inspection, Measurement, and Test coverage for the Inspection factor.

You can focus on one area at a time to further examine the possible causes or sources of variation for each major factor.

Prepare the Data

Before you produce the diagram, begin with your data in two columns of a data table.

Figure 13.3 Example of the Ishikawa.jmp Data Table

	Parent	Child
1	Defects in circuit board	Inspection
2	Defects in circuit board	Solder process
3	Defects in circuit board	Raw card
4	Defects in circuit board	Components
5	Defects in circuit board	Component insertion
6	Inspection	Measurement
7	Inspection	Test coverage
8	Inspection	Inspector
9	Solder process	Splatter
10	Solder process	Flux
11	Solder process	Chain speed
12	Solder process	Temperature
13	Solder process	Wave pump
14	Temperature	Setup

Notice that the Parent value Defects in circuit board (the effect) has five major factors, listed in the Child column. One of these major factors is Inspection, which has its own causes listed in the Child column. Parent values have children, and children can have their own children (and therefore be listed in both the Parent and Child columns.)

Launch the Diagram Platform

Launch the Diagram platform by selecting **Analyze > Quality And Process > Diagram**.

Figure 13.4 The Diagram Launch Window

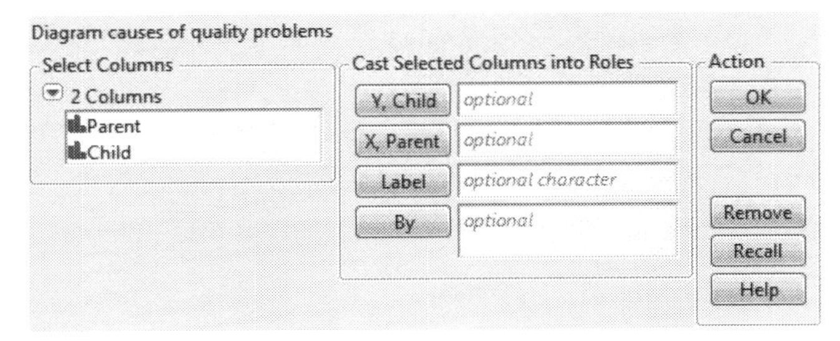

Tip: To create a basic diagram that is not based on a data table, leave the **Y, Child**, and **X, Parent** fields empty and click **OK**. Then edit the nodes using the options in the right-click menu. See "Right-Click Menus" on page 306.

Y, Child Represents the child factors contributing to the parent factors.

X, Parent Represents the parent factors (including the effect) that have child factors.

Label Includes the text from the Label columns in the nodes of the diagram.

By Produces separate diagrams for each value of the By variable.

The Cause-and-Effect Diagram

Figure 13.5 Cause-and-Effect Diagram

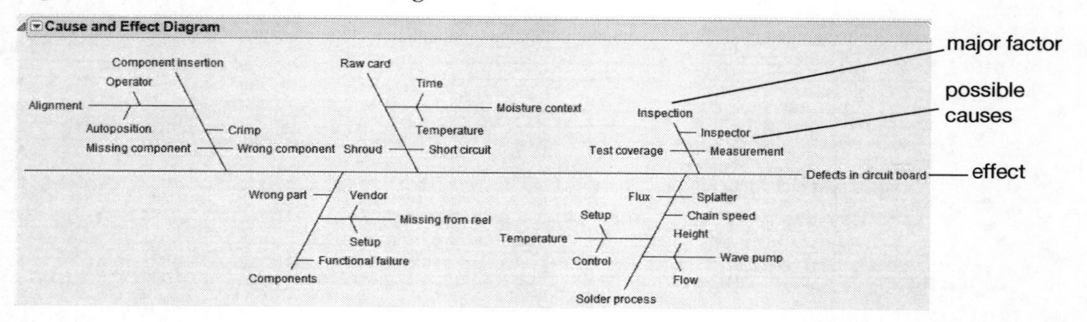

In Figure 13.5, the effect or problem, Defects in circuit board, appears on the right as the center line. The major contributing factors appear at the end of the branches (Inspection, Solder process, Raw Card, and so on.) Possible causes branch off each major factor.

Right-Click Menus

Right-click on a highlighted node to modify text, insert new nodes, change the diagram type, and more. Note the following:

- Right-click on a title to change the font and color, positioning, visibility, or formatting.
- Click and highlight a node to rename it.
- Click and drag a node to move it.

Text Menu

The Text menu contains the following options:

Font Select the font of the text or numeric characters.

Color Select the color of the text or numeric characters.

Rotate Left, Rotate Right, Horizontal Rotates the text or numbers to be horizontal, 90 degrees left, or 90 degrees right.

Insert Menu

Use the **Insert** menu to insert items onto existing nodes. The Insert menu contains the following options:

Before Inserts a new node to the right of the highlighted node. For example, Figure 13.6 inserts Child 1.5 before Child 2.

Figure 13.6 Insert Before

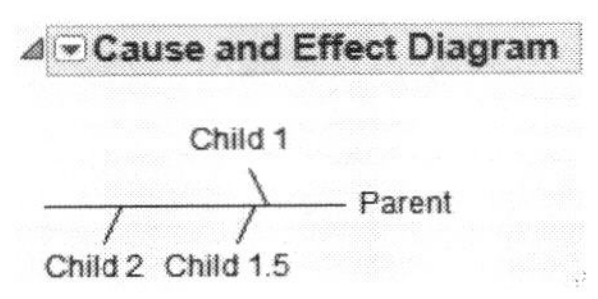

After Inserts a new node to the left of the highlighted node. For example, Figure 13.7 inserts Child 3 after Child 2.

Figure 13.7 Insert After

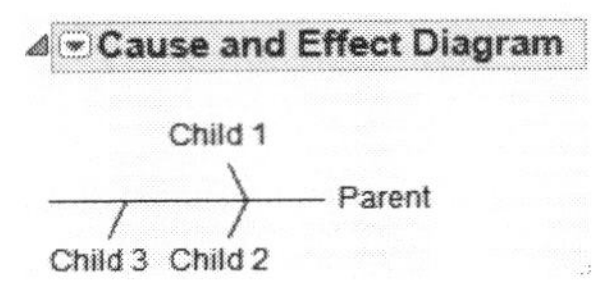

Above Inserts a new node at a level above the current node. For example, Figure 13.8 inserts Grandparent at a level above Parent.

Figure 13.8 Insert Above

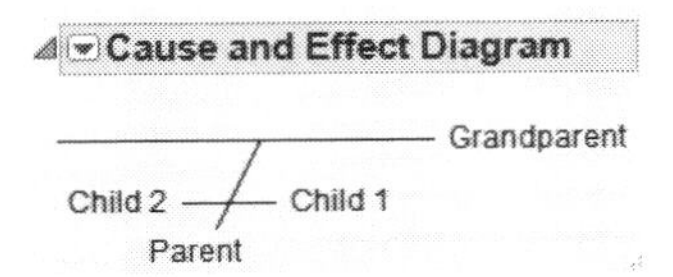

Below Inserts a new node at a level below the current node. For example, Figure 13.9 inserts Grandchild at a level below Child 2.

Figure 13.9 Insert Below

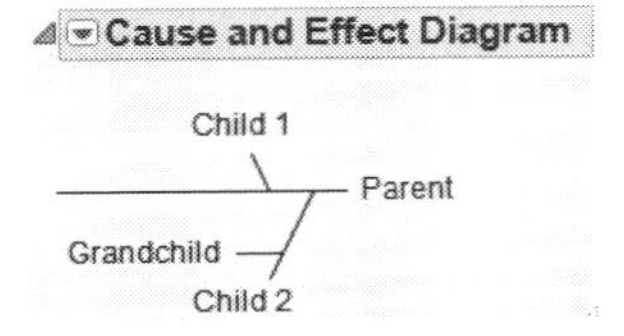

Move Menu

Use the Move menu to move nodes or branches. The Move menu contains the following options:

First Moves the highlighted node to the first position under its parent.

Last Moves the highlighted node to the last position under its parent.

Other Side Moves the highlighted node to the opposite side of its parent line.

Force Left Makes all horizontally drawn elements appear to the left of their parent.

Force Right Makes all horizontally drawn elements appear to the right of their parent.

Force Up Makes all vertically drawn elements appear above their parent.

Force Down Makes all vertically drawn elements appear below their parent.

Force Alternate Draws children on alternate sides of the parent line.

Figure 13.10 Force Options

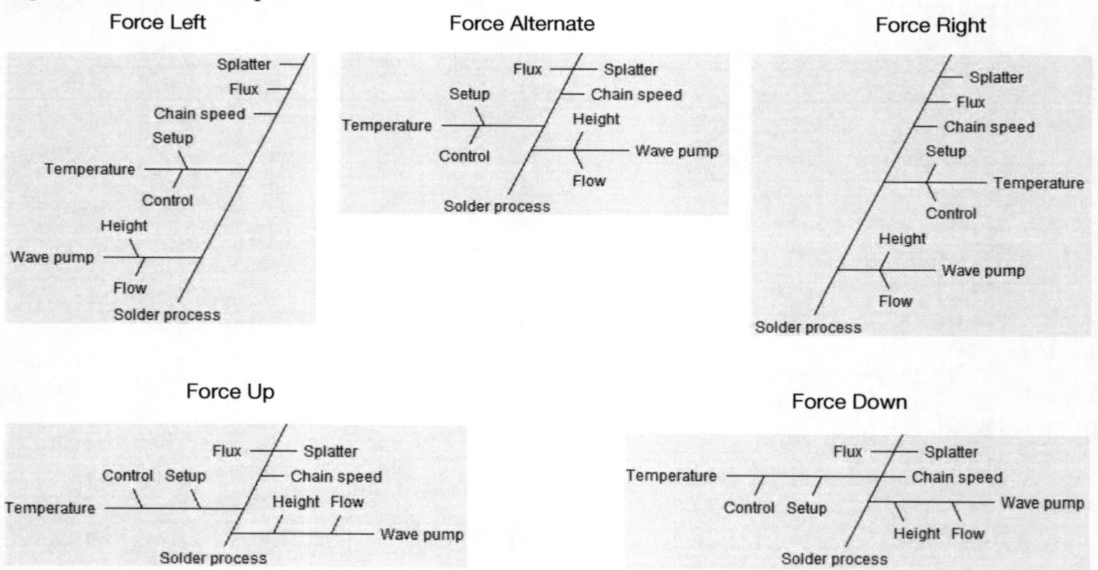

Other Menu Options

The right-click menu for a highlighted node also contains these options:

Change Type Changes the entire chart type to **Fishbone, Hierarchy,** or **Nested.**

Uneditable Disables all other commands except **Move** and **Change Type.**

Text Wrap Width Specifies the width of labels where text wrapping occurs.

Make Into Data Table Converts the currently highlighted node into a data table. Convert the all nodes by highlighting the whole diagram (effect).

Close Shows the highlighted node.

Delete Deletes the highlighted node and all of its children.

Save the Diagram

There are different ways to save your diagram. Choose from one of the following:

- save the diagram as a data table
- save the diagram as a journal
- save the diagram as a script

Save the Diagram as a Data Table

1. Highlight the entire diagram.
2. Right-click and select **Make Into Data Table**.
3. Save the new data table.

Note the following about this approach:

- If you have other processes that need to update the data table, this can be a good approach to choose.
- Very little customization is available, because the data table cannot represent the customization.

Save the Diagram as a Journal

1. Highlight the entire diagram.
2. Right-click and select **Edit > Journal**.
3. Save the new journal.

Note the following about this approach:

- This option can be a good choice for impromptu work. For example, you can manually build the diagram, save it as a journal, then reopen the journal later and continue building and editing the diagram.
- Any customization exists only in the journal, and the journal is not connected to the data table.

Save the Diagram as a Script

1. From the red triangle menu, select **Save Script > To Script Window**.
2. Save the new script.

Note the following about this approach:

- If you have other processes that need to update the data table, this can be a good approach to choose.
- If you created the diagram from a data table, a simple script appears that relaunches against the data table with no customization.
- If you created the diagram without using a data table (or from a journal), a more complex script appears that contains all the commands needed to add and customize each area of the diagram.

Appendix A

References

Agresti, A., and Coull, B. (1998), "Approximate is Better Than 'Exact' for Interval Estimation of Binomial Proportions," *The American Statistician*, 52, 119–126

American Society for Quality Statistics Division (2004), *Glossary and Tables for Statistical Quality Control*, Fourth Edition, Milwaukee: Quality Press.

Automotive Industry Action Group (AIAG) (2002), *Measurement Systems Analysis Reference Manual*, Third Edition.

Automotive Industry Action Group (AIAG) (2005), *Statistical Process Control*, Second Edition.

Barrentine (1991), *Concepts for R&R Studies*, Milwaukee, WI: ASQC Quality Press.

Bothe, D.R. (1997), *Measuring Process Capability: Techniques and Calculations for Quality and Manufacturing Engineers*, McGraw-Hill.

Fleiss, J. L. (1981). Statistical Methods for Rates and Proportions. New York: John Wiley and Sons.

Kourti, T. and MacGregor, J. F. (1996), "Multivariate SPC Methods for Process and Product Monitoring," Journal of Quality Technology, 28:4, 409-428.

Lucas, J.M. (1976), "The Design and Use of V–Mask Control Schemes," *Journal of Quality Technology*, 8, 1–12.

Lucas, J.M. and Crosier, R.B. (1982), "Fast Initial Response for CUSUM Quality Control Schemes: Give Your CUSUM a Head Start," *Technometrics*, 24, 199-205.

Meeker, W. Q. and Escobar, L. A., (1998), *Statistical Methods for Reliability Data*, John Wiley & Sons.

Montgomery, D.C. (2013), *Introduction to Statistical Quality Control*, 7th Edition New York: John Wiley and Sons.

Nair, V.N. (1984), "Confidence Bands for Survival Functions with Censored Data: A Comparative Study," *Technometrics*, 26:3, 265-275.

Nelson, L. (1984), "The Shewhart Control Chart—Tests for Special Causes," *Journal of Quality Technology*, 15, 237–239.

Nelson, L. (1985), "Interpreting Shewhart X Control Charts," *Journal of Quality Technology*, 17, 114–116.

Nelson, W.B. (1982), *Applied Life Data Analysis*, New York: John Wiley and Sons.

Portnoy, S.(1971), "Formal Bayes Estimation with Application to a Random Effects Model", *The Annals of Mathematical Statistics*, Vol. 42, No. 4, pp. 1379–1402.

Sahai, H. (1974), "Some Formal Bayes Estimators of Variance Components in the Balanced Three-Stage Nested Random Effects Model", *Communication in Statistics – Simulation and Computation*, 3:3, 233–242.

Slifker, J.F., Shapiro, S.S. (1980), "The Johnson System: Selection and Parameter Estimation," *Technometrics*, 22, 239-246.

Sullivan, J.H. and Woodall, W.H. (2000), "Change-point detection of mean vector or covariance matrix shifts using multivariate individual observations," *IIE Transactions*, 32, 537-549.

Tracy, N. D., Young, J. C. and Mason, R. L. (1992), "Multivariate Control Charts for Individual Observations," Journal of Quality Technology, 24:2, 88-95.

Westgard, J.O. (2002), *Basic QC Practices, 2nd Edition*. Madison, Wisconsin: Westgard QC Inc.

Wheeler, Donald J. (2004) *Advanced Topics in Statistical Process Control*, 2nd Edition. SPC Press.

Wheeler, Donald J. (2006) *EMP III Using Imperfect Data*. SPC Press.

Wludyka, P. and Sa, P. (2004) "A robust I-Sample analysis of means type randomization test for variances for unbalanced designs," *Journal of Statistical Computation and Simulation*, 74:10,701-726

Index

Quality and Process Methods

Symbols

A

B

C

D

E

F

G

I

K

L

M

N

O

P

W-Z

Made in the USA
San Bernardino, CA
08 February 2017